Advanced Ceramic Membranes and Applications

T0225315

Advanced Ceramic Membranes and Applications

Chandan Das
Sujoy Bose

CRC Press

Taylor & Francis Group
Boca Raton London New York

CRC Press is an imprint of the
Taylor & Francis Group, an **informa** business

CRC Press
Taylor & Francis Group
6000 Broken Sound Parkway NW, Suite 300
Boca Raton, FL 33487-2742

First issued in paperback 2020

© 2017 by Taylor & Francis Group, LLC
CRC Press is an imprint of Taylor & Francis Group, an Informa business

No claim to original U.S. Government works

ISBN-13: 978-0-367-57318-8 (pbk)
ISBN-13: 978-1-138-05540-7 (hbk)

Visit the Taylor & Francis Web site at
http://www.taylorandfrancis.com

and the CRC Press Web site at
http://www.crcpress.com

Contents

Preface

Ceramic membranes have been established as very effective but commercially not very attractive options for separation and purification processes in chemical and allied industries dealing with petrochemicals and chemicals, food and beverages, biochemical removal of different types of pollutants, and so on, due to a smaller amount of economic feasibility. Ceramic membranes are now competitive with traditional techniques because they are benign in the environment, and they can provide high mechanical, thermal, and chemical stability, though they require relatively high manufacturing and investment costs. From this standpoint, scientists and engineers are working hard and approaching the innovation of ceramic membranes because of low-cost raw materials that allow the fabrication of ceramic membranes.

This book is based upon the basic fundamental concepts of ceramic membranes, as well as practical and theoretical knowledge of conventional and advanced ceramic membranes combined with unorthodox ideas for novel approaches to them. Also covered is the research work published or presented by the Membrane and Bioremediation Lab, Department of Chemical Engineering, IIT Guwahati, over the last few years by both the authors. A selection of these works is included, along with the fundamental concepts of ceramic membranes. Some of the knowledge gained from the authors' present research work is used as practical examples in the book. We heartily regret that it was not possible to take in all the work done at the laboratory during that time. Still, we hope that most readers consider the selection of work presented along with the theoretical concept of ceramics in this book satisfactory. Moreover, we hope that at least some of our readers agree with us that it is a good time to carry out steps toward advanced ceramic membranes with a novel approach going beyond the conventional methodologies treating reactions and separation simultaneously. That is why we deliberately included the concept of catalytic membranes and membrane reactors.

A good number of books are available on the topic and no doubt many of them are of excellent quality within their scope. This book is a humble effort to provide a balanced blend of fundamental concepts of fabrication, characterization of conventional ceramics, and recent advances in the field of ceramic membranes. On the one hand, this book is a guideline for students, scientists, and engineers, and on the other, it provides new ideas for creative thinkers. It is a view into the past, the present, and the future.

This book is organized and divided into nine chapters. Basically, we have started with the conventional theories of ceramic membranes, proceeded with new and novel ideas, and ended with advanced uses of ceramic membrane, as well as advances in membrane fabrication techniques. For readers new to this field, Chapter 1 provides an overview of different types of ceramic membranes, such as microfiltration (MF), ultrafiltration (UF), nanofiltration (NF), gas separation, and pervaporation, as well as recent developments in the field of ceramic membranes. Chapter 2 deals with the fabrication processes of ceramic membranes and factors such as raw materials, pore-formers, and sintering that are associated with them. This chapter also describes different shapes, modules, and flow patterns suitable for ceramic membranes. Chapter 3 focuses on the different characterization techniques for ceramic membranes. After extensive and continuous activity (preferably dealing with highly viscous fluid), membrane efficiency can be reduced drastically by the action of pore blocking and concentration polarization. Hence, it is necessary to have a clear idea about enhancing the lifetime of a ceramic membrane by minimizing the factors mentioned earlier. For this perspective, we included an entire chapter on the concept of ceramic membrane cleaning techniques, as detailed in Chapter 4. Chapter 5 is devoted to the fundamentals of newly developed advanced ceramic membranes and membrane reactors dealing with different catalytic reactions associated with petrochemical industries. This chapter also explains the classifications of membrane reactors and provides an idea on how they differ from conventional ones. Chapters 6 and 7 explain the fabrication of catalytic membranes and catalytic membrane reactors (CMRs). Furthermore, a practical example is given to understand the approach to fabricating CMRs. Chapter 8 describes the effect of mass transfer on the activity of the CMR. Chapter 9 elaborates various applications of ceramic membranes such as removal of volatile organic compounds, acid gases, SO_2, and mercury; wastewater treatment, fruit juice clarification, and heavy metal separation. Furthermore, examples of membrane contactors for the recovery of aroma compounds are presented. A novel application of a ceramic membrane in a fuel cell is also described in this chapter. The commercial applications along with other applications, including new categories of use for ceramic membranes, are also discussed.

    We wish to thank the Central Instrumental Facility, IIT Guwahati, for conducting all the experiments for characterizing the catalytic membrane. We would like to thank the entire departmental nonteaching lab staff for their enormous support. Moreover, Dr. Bose would like to thank his parents and family members and all the well-wishers for their constant support.

Chandan Das and Sujoy Bose
Guwahati and Kozhikode

Acknowledgments

We would like to express our gratitude to all those who helped and guided us, directly or indirectly, in different ways to completing this book within the time span of 1 year. First of all, we would like to express our sincere gratitude to Dr. Gagandeep Singh, senior editor, Taylor & Francis, for enabling us to publish this book. We are indebted to him for his useful suggestions and necessary arrangements made for us throughout the entire period.

We are grateful to all those who provided support, talked things over, read, wrote, offered comments, allowed us to quote their remarks, and assisted in the editing, proofreading, and design.

We must also thank the faculty members of the Department of Chemical Engineering, IIT Guwahati, for their kind cooperation during our project tenure in the department. We are also thankful to all staff members and laboratory superintendents, especially Dr. Lukumoni Borah, Mr. Kaustavmani Deka, and Mr. Dipak Kumar Barman of the Chemical Engineering Department, IIT Guwahati, for their generous help during our entire research period.

We are also grateful to all our colleagues who supported and encouraged us continuously.

We are thankful to Prof. Pallab Ghosh for his valuable suggestion in writing the book. We should acknowledge the support provided by Prof. Bishnupada Mandal, HOD, Department of Chemical Engineering, IIT Guwahati, and Prof. V. Sivasubramanian, Department of Chemical Engineering, NIT Calicut.

We are also thankful to the Central Instruments Facility of IIT Guwahati for allowing us to carry out scanning electron microscopy analysis, field emission scanning electron microscopy analysis, electron spin resonance, surface area analyzer, and Raman spectroscopy analysis, which were vital in this research work. In this regard, we should acknowledge the assistance provided by Dr. Kula Kamala Senapati; Mr. Chandan Borgohain, scientific officer; and Mr. Madhurjya Borah, junior technical superintendent, Central Instruments Facility, IIT Guwahati, and all the operators allotted for these instruments.

We are also thankful to the central workshop of IIT Guwahati for helping us in the fabrication of our experimental setup, which was the most essential part of the research work given as examples in the book.

We were fortunate enough to get excellent people like Dr. Mahesh Kumar Gagrai, Dr. Vijay Singh, Dr. Arijit Das, Mr. Suman Saha, Mr. Kibrom Alebel Gebru, Mr. Amit Baran Das, Mr. Abhisek Shukla, Mr. Abhradip Pal, Mr. Chelli V. Rao, and Mr. Raj Kumar Das for their continuous support. Special thanks to Mr. Rishiket Kundu, Mr. Vishesh Dhaliwal, Mr. Rahul Dohare, Mr. Vineet

Kumar, and Ms. Shubhangi Parde for their help and cooperation in our research work.

Most of all, we would like to express our deepest sense of gratitude to all our family members, our parents, and all well-wishers. Their love, care, sacrifices, and encouragement have made it possible for us to come so far.

Authors

Dr. Chandan Das received his bachelor's and master's degrees in chemical engineering from the University of Calcutta and his PhD in chemical engineering from IIT Kharagpur; he is currently an associate professor in the Chemical Engineering Department at IIT Guwahati. Thus far, he has guided four doctoral degree students and eighteen students studying for master's degrees in technology; he is currently guiding eight PhD candidates. Dr. Das was the recipient of the "Dr. A.V. Rama Rao Foundation's Best PhD Thesis and Research Award in Chemical Engineering/Technology" in 2010 from the Indian Institute of Chemical Engineers (IIChE). Dr. Das has authored about one hundred technical publications in peer-reviewed journals and proceedings. He has authored one book, *Treatment of Tannery Effluent by Membrane Separation Technology*, published by Nova Science Publishers, Hauppauge, New York, and contributed two book chapters to other books. He has two patents to his credit. He has handled five sponsored projects and five consultancy projects and has visited Denmark, Malaysia, Sri Lanka, and Japan to exchange ideas.

Because of his association with various research work in the area of water and wastewater treatment, such as treatment of tannery wastewater using membrane separation technology, as well as removal of pollutants using micellar-enhanced ultrafiltration, Dr. Das has gained expertise in membrane separation technology for removing various pollutants from contaminated water and wastewater. His research activity encompasses understanding of fundamental principles during filtration as well as the development of technology based on membrane separation. In particular, his research areas are modeling of microfiltration, ultrafiltration, nanofiltration, reverse osmosis, treatment of oily wastewater, and tannery effluent using membrane-based processes. He is exploring the detailed quantification of flux decline from fundamentals. As an offshoot of this major research, he has fabricated ceramic membranes using low-cost precursors such as sawdust.

Dr. Das is also working on decontamination of chromium-laden aqueous effluent using *Spirulina platensis*. He is actively involved in the production of high-value-added products—namely, total phenolics, flavonoids, tocopherol, etc., from black rice as well as 6-gingerol, vitamin C content, and essential oil content from ginger in northeast India.

Dr. Das's achievements include:

1. Separation of Cr(III) from aqueous solution by sorption on *Spirulina platensis* at pH 6.2. Phosphate, carboxylic, and amine groups are responsible for metal ion binding. The phosphate group has maximum attachment with ions—dead: (adsorption + ppt) 100% (feed: 20 ppm); live: 97% (feed: 100 ppm)

2. Cr(VI) reduction into Cr(III) at acidic pH using protonated functional groups; 98% reduction at 0.5 pH; temp-high reduction high

3. PCTE in ZMC reasonable *Spirulina* cell growth

4. Cross-flow microfiltration of oily wastewater retains 98% oil droplets; COD reduction from 2000 to 225 ppm, combined diffusion model (0.01–7 µm)

5. Ceramic membrane suitable for N-methyl-2-pyrrolidone (NMP) separation from coal–NMP mixture; tested up to the third stage; iron ore slime, steel industry waste utilized for ceramic membrane fabrication

6. Through a spiral-wound membrane module, the with-permeate recycle had better quality compared to the without-permeate recycle; laminar Brownian diffusion model (<0.1 µm); turbulent shear induced (0.5–30 µm) due to clustering

Dr. Das's major research fields include:

- Field 1: membrane separation technology: Dr. Vijay Singh, Dr. Sujoy Bose, Mr. Suman Saha, Mr. Kibrom Alebel Gebru, Mr. Kulbhushan Samal
    - Membrane modules: (1) unstirred batch cell; (2) stirred batch cell; (3) cross-flow cell; (4) spiral wound; (5) spinning basket; (6) tubular (catalytic) membrane module; (7) electrospun nanofiber membrane
    - Applications: oily wastewater, coal-solvent mixture, fruit juice, elemental S, water and wastewater, humic acids
- Field 2: bioremediation: Dr. Mahesh Kumar Gagrai
    - *Spirulina platensis*, blue-green microalgae: removal of Cr(III) and Cr(VI) from aqueous solution
- Field 3: extraction, purification, and separation of value-added products from herbs and crops: Dr. Arijit Das, Mr. Abhishek Shukla, Mr. Amit Baran Das
    - Aloe vera, rebaudioside (RA), aloe vara, ginger, black and red rice of the northeastern region of India

 **Dr. Sujoy Bose** is an assistant professor in the Department of Chemical Engineering, National Institute of Technology, Calicut. He graduated from the Durgapur Institute of Advanced Technology and Management, West Bengal University of Technology. He finished postgraduate work at the National Institute of Technology, Durgapur, and received his PhD from the Indian Institute of Technology, Guwahati. He has over 7 years of experience in the field of membrane technology. He is an expert in the area of developing ceramic membranes for different applications. Recently, he introduced sawdust as a novel, inexpensive, raw material for ceramic membranes. His research interests include materials science and nanofluids. He has published in peer-reviewed international journals such as *Materials Letters, Ceramics International*, and *Industrial and Engineering Chemistry Research*. He served as an executive committee member of the IIChE, Guwahati region chapter, from 2013 to 2014.

1

Ceramic Membrane Processes

1.1 Introduction to Ceramic Membranes

We start this book with a brief idea about the basics and fundamental concepts of ceramic membranes and their recent trends in different drives along with examples. An overview of the fabrication, characterization, and use of low-cost ceramic membranes in the development of hybrid membranes is then discussed in the following chapters; then we move to the most challenging and advanced sector of membrane technology (i.e., catalytic membrane and membrane reactors and their applications).

In general, a ceramic membrane can be defined as a permselective barrier placed between two phases that permits one or more components to selectively pass from one phase to the other in the presence of an appropriate driving force.

In the fourteenth century, ceramic was mainly known as an art and used widely for interior home decorations (pottery, tableware, and cookware) and still is [1]. Any details of ceramic processing and manufacturing were difficult to understand as researchers showed no interest. But, from the nineteenth century on, the scenario has completely changed. Researchers have started showing interest in ceramic material in the field of novel separations due to its high thermal resistivity, excellent mechanical and chemical stability, and, most importantly, high permeability and selectivity as performance parameters. Ceramics now include domestic, industrial, and building products, as well as a wide range of ceramic art and are in a strong position to compete with polymeric membranes. From the twentieth century, new ceramic materials have been developed for use in advanced ceramic engineering, such as structural ceramics, electrical and electronic ceramics, ceramic coating and chemical processing, and environmental ceramics [2].

Ceramic membranes are usually porous and dense in nature. Factors like permeation and separation for porous ceramic membranes are based on thickness, pore size, and porosity of the membrane, whereas, for dense ceramic membranes, the concept of permeation and separation is more difficult to predict. Applications and separations in porous and dense ceramic membranes are mainly based on their pore size as shown in Table 1.1.

TABLE 1.1

Classification of Porous Ceramic Membranes

Category	Pore Size (nm)	Separation Mechanism	Applications
Microporous	<2	Micropore diffusion	Gas separation
Mesoporous	2–50	Knudsen diffusion	UF, NF, gas separation
Macroporous	>50	Sieving	UF, MF
Dense	–	Diffusion	Gas separation

Notes: MF = microfiltration; NF = nanofiltration; UF = ultrafiltration.

Ceramic membranes are typically a combination of a number of layers of one or more different materials by conventional means, called a composite. These layers are categorized in three different segments as macroporous support (bottom layer), mesoporous intermediate layers (one or more), and a microporous top surface layer, as shown in Figure 1.1. The bottom layer (macroporous support) provides mechanical support, whereas the intermediate layers connect the pore size differences between the support layer and the top layer; the actual separation takes place at the top layer [3].

Manufacturing of ceramic membranes is usually based on many different compositions combined with pore-formers and binders in a wide variety of proportions. Based on a literature survey, the use of raw materials, including pore-formers and binders in terms of manufacturing cost and its effect on properties, is an influential and effective approach. This attention to the raw materials along with pore-formers and binders and its influence on membrane morphology (pore size, porosity, and surface texture) and thermal, mechanical, and chemical stability is the principal concept of research on ceramic membranes.

To be completely successful, the pore-forming mechanism of pore-formers must be understood in a broad sense. On the other hand, we are concerned about the influence of pore-formers on the morphology and mechanical and chemical stability of the ceramic membrane. Equally important is the way in which binders like sodium metasilicate, boric acid, etc., are reacted with other raw materials during sintering in the fabrication of ceramic membrane,

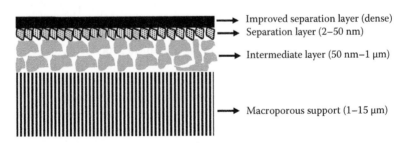

FIGURE 1.1
Schematic diagram of ceramic membrane (composite).

by means of not only the mechanical strength but also the interparticle distance and surface matrix defects.

Usually, the atoms or ions in ceramics are arranged in crystalline as well as amorphous (glass) phases. The arrangement of phases (i.e., crystalline and amorphous) is often controlled by porosity of the ceramic. A change in percent porosity from higher (porous) to lower (dense) makes a ceramic apparent (crystalline) in nature and vice versa. A change in pore size modifies a ceramic membrane from dense to permeable (porous). Properties of a ceramic such as thermal conductivity or electrical conductivity are influenced by grain boundaries. A grain boundary is the border between two grains, or crystallites, in a polycrystalline material. Basically, grain boundaries are flaws in the crystal structure. A decrease in grain boundary may reduce thermal and electrical conductivity. Furthermore, a decrease in grain size may change a ceramic from feeble and soft to durable and rigid. An amorphous ceramic withstands deformation at higher temperatures, while a crystalline one deforms readily. The change in phase arrangement can also change a ceramic from an insulator to a conductor. It should also be understood that some materials experience changes in their electrical properties under different conditions. Glass, for instance, is a very good insulator at room temperature, but becomes a conductor when heated to a very high temperature. With increase in temperature, molecular arrangement changes due to high vibrational energy, causing change in interfaces between atomic layers, and is responsible for minimizing movement of thermal resistances between these layers. In addition, the change in phase of ceramic by suitable heat treatment (sintering) can effectively alter many of its properties and increase or decrease its performance efficiency.

These interpretations are of theoretical interest. We can now become even more focused toward innovative ideas that not only offer minimum fabrication cost of ceramic membrane but also provide high performance. This new approach provides us more information about novel and cheap raw materials, including pore-former compositions. Moreover, it gives the basis for understanding the change in membrane properties such as porosity, pore size, surface texture, etc.

1.2 Microfiltration

Microfiltration (MF) is a low-pressure (less than 2 bar) driven membrane process for separating colloids, suspended particles, bacteria, and yeast cells from a liquid or particulates from a gas using a ceramic porous membrane on the basis of a sieving mechanism. The ceramic membranes used in microfiltration have pore sizes in the range between 0.05 and 10 μm, as shown in Figure 1.2.

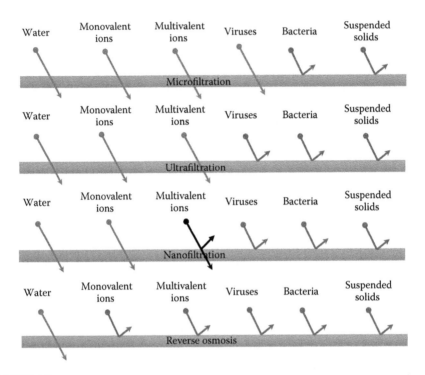

FIGURE 1.2
Pore size range for ceramic membranes.

The ceramic MF membranes are made from metal oxides (alumina), glass, zirconia, titania, silica (quartz), clays (rocks, kaolinite) and natural clays, etc. Selection of raw materials is directly reflected in their physical, mechanical, chemical, and morphological properties as well as applications and manufacturing cost. Microfiltration membranes have also been developed from a wide range of polymers like cellulose nitrate, cellulose acetate, polysulfone, polyvinyl alcohol, polyvinylidene fluoride, etc., but are less attractive compared to ceramic membranes due to thermal irresistibility, mainly because the melting point of polymers lies below 200°C.

MF membranes have a large range of pores, which causes transportation of solvent through the pores by convection. This can be easily described by the Hagen-Poiseuille equation, if the nature of the pores is assumed to be cylindrical. Microfiltration membranes are generally prepared as a film supported over a macroporous ceramic support, which gives excellent mechanical strength, but offers negligible gas transfer resistance during separation of gas. Sometimes, an intermediate film is introduced between the ceramic support and the top separation film to reduce the opening between large pores of the support and the small pores of the top separation film. MF ceramic membranes possess two different types of pore structures: tortuous and capillary pores. Tortuous pore membranes provide spongy structure, high surface area, long life, and high dirt-loading capacity. The capillary pore membrane offers a cylinder-like structure with well-defined pore size and pore structure.

The transport mechanism for liquids can be described on hydrodynamic flow of solvent and slow diffusion of solutes. For transport of gases through the pores of ceramic membranes, these are assumed to be cylindrical like a capillary tube or finger-like structure and can be defined by the kinetic theory of gases. Based on the relative magnitudes of the pore radius, r_p, and mean free path, λ, the gas molecules pass through capillaries by any one of the three proposed transport mechanisms—namely, Knudsen flow, slip flow, and viscous flow. Knudsen type flow is predominant when the mean free path of the gas molecules is much greater than the pore size of the membrane (i.e., $(r_p/\lambda) < 0.05$ [4]), whereas slip flow occurs in the range $(r_p/\lambda) = 0.05$ to 3 and viscous flow is considered when $(r_p/\lambda) > 3$ [5]. The mean free path of a gas molecule is known as

$$\lambda = \frac{RT}{\sqrt{2}\pi d^2 N \bar{p}} \tag{1.1}$$

where
  R is the gas constant
  8.314 m^3 Pa K^{-1} mol^{-1}
  T is the temperature in K
  d is the collision diameter of the gas molecules in m
  N is the Avogadro number
  $\bar{p}$, the average pressure across the membrane, is defined as $\bar{p} = (p_h + p_1)/2$, where p_h and p_1 are the upstream and downstream pressures, respectively, in Pa

In Knudsen flow, the collisions of the molecules with the wall of the membrane pore is more repeated than the collisions among molecules. The flux due to Knudsen flow is described by the Knudsen equation [6] in terms of concentration or partial pressure of gas molecules:

$$J_K = -D_K \frac{dC}{dz} \tag{1.2}$$

or

$$J_K = -D_K \frac{dp}{dz} \tag{1.3}$$

where D_K is the Knudsen diffusion coefficient, defined as

$$D_K = \frac{2}{3} u r_p \tag{1.4}$$

where u, the mean molecular speed, can be obtained from the kinetic theory of gases [7,8]:

$$u = \sqrt{\frac{8RT}{\pi M}} \tag{1.5}$$

Again, M is the molecular weight of the gaseous species.

The slip flow in a porous ceramic membrane is described as the modified version of Knudsen flow and is given as

$$D_K = \frac{2}{3} \frac{\varepsilon}{\tau} u r_p \tag{1.6}$$

(Equation 1.6 is a modified version of Equation 1.4.)

where τ is the tortuosity factor of the ceramic membrane, and ε is the porosity of the membrane.

Viscous flow occurs due to molecule–molecule collision, when the mean free path is much smaller than the pore size of the membrane. Mathematically, it is defined as

$$J_V = -\frac{r_p^2}{8\mu} \frac{p}{RT} \left(\frac{dp}{dz} \right) \tag{1.7}$$

MF ceramic membranes are applied in several applications, such as sterilization and clarification of fruit juices like mosambi juice [9] and orange juice [10], treatment of oily wastewater [11], toxic brilliant green dye from aqueous medium [12], separation of coal from organic solvent [13], etc.

1.3 Ultrafiltration

Ultrafiltration (UF) is a pressure-driven membrane separation process where porous membranes are used to separate relatively large molecules (the approximate range of molecular weight is from 10^{-3} to 80×10^{-3} Da) from smaller ones or a colloidal suspension. The mechanism for separation of the solvent from the solute is called sieving or size exclusion. A sieving mechanism is a process by which the rejection of solute is determined based on the pore size and pore size distribution of the ceramic membranes. The purification efficiency of a porous ceramic membrane also depends on surface interactions between the membrane surface and solutes/solvents. In UF, low applied pressures (1–10 bar) are sufficient to achieve high permeability (1–10 $m^3.m^{-2}.day^{-1}.bar^{-1}$). Usually, pore diameter in UF membranes ranges between 1 and 100 nm.

UF membranes have been prepared from a wide range of polymer materials (polysulfone, polyacrilonitrile, cellulose acetate, aromatic polyamides, polyvinyl alcohol, polycarbonate, etc.) as well as ceramic materials such as alumina, titania, and zirconia. The growth of ceramic membranes for UF applications compared to polymeric ones is mainly due to high thermal, chemical, and mechanical resistivity.

Separation efficiency of a UF membrane depends on several factors, such as operating pressure, transmembrane flow, operating temperature, salt concentration, etc. Permeation rate is directly proportional to the applied pressure, flow velocity, and temperature across the ceramic membrane surface. Sometimes, due to enhanced fouling and pore blockage, the operating pressures exceed 10 bar. Fouling inside the pores of a membrane can also be reduced by increasing the flow velocity. Normally, in a laminar flow regime, flow velocity is maintained at 1 to 2 $m.s^{-1}$ in thin channel membrane modules. However, in tubular ceramic membrane modules, flow velocity may be kept up to 5 $m.s^{-1}$. The effect of temperature on membrane flux is important in order to discriminate between permeate declination and the effect of other parameters. With rising temperature, viscosity of retentate decreases and diffusivity of macromolecules increases; thus permeability for a porous ceramic membrane increases as well. However, the ultrafiltration rate decreases with increasing molar concentration. UF membranes possibly will not retain the salts, but due to their ionic nature, some intramolecular association takes place that eventually reduces the UF rate. UF permits the removal of up to 90% of water at ambient temperature.

One of the main problems of the UF process is membrane fouling caused by enhanced solute concentration at the membrane surface, which leads to concentration polarization. Several techniques to avoid or prevent fouling include cleaning the membranes, pretreating the feed water, using tangential and frequent feed flow, mixing promoters into the feed channel, introducing pressure pulses of the feed flow, and adding dynamic particles to the feed stream.

Typical applications of UF processes using ceramic membranes are found in paint, pulp and paper, textile, dairy, food, pharmaceutical, and biological industries and water treatment plants. A detailed review on different applications of ceramic membranes can be found elsewhere [14].

1.4 Nanofiltration

Nanofiltration (NF) is a high-pressure driven membrane separation process, approximately 4–10 bar. Similarly to reverse osmosis (RO; see following section), it allows selective passage of solvent while solutes are retained partially or completely. NF membranes can only reject multivalent ions with negligible selectivity toward the monovalent ion, which is the only difference from the RO technique.

In the past, researchers have shown interest in polymeric NF membranes only for seawater desalination, whey production, recovery of dyes, separation of heavy metals from acid solution, etc. But, major problems associated with NF membranes, such as excessive fouling, low resistance to chlorine and different oxidants, extensive usage of chemical for pretreatment, less longevity, and low thermal and mechanical stability, lower popularity compared to ceramic membranes. However, the high fabricating cost and low packing density make commercially available ceramic membranes economically unfeasible and unrealistic for different applications. But, nowadays, research in the development of ceramic NF membranes has been carried out extensively, and membranes prepared from graphene oxide [15], zirconia [16–19], silica–zirconia [20,21], titania [22–25], γ-alumina [26–29], palladium–titania [30], and α-alumina [31] have been reported. All these membranes have been fabricated for water and wastewater treatment using a sol-gel colloidal process.

1.5 Reverse Osmosis

Reverse osmosis is a high-pressure, energy-efficient method with a pore size <2 nm for desalination of seawater and brackish water, wastewater treatment, and ultrapure water production; it is also widely considered for polymeric membranes like NF membranes. Usually, there are no pores in RO membranes in the physical sense. Pores in RO are the channel of the solvent, and solute molecules occur through the spaces in the polymer matrix. The driving force for RO membranes is the difference in pressure for solvent transport and concentration difference for solute flow—specifically, the difference in chemical potential.

As shown in Figure 1.3, osmosis is a natural phenomenon where an aqueous solution of a substance is kept detached from water using a membrane (semipermeable) in a two-compartment cell. Water diffuses through the membrane into the higher concentration cell (Figure 1.3a).

This situation happens due to potential chemical difference of water between the two cells. In the normal osmosis process, the solvent naturally

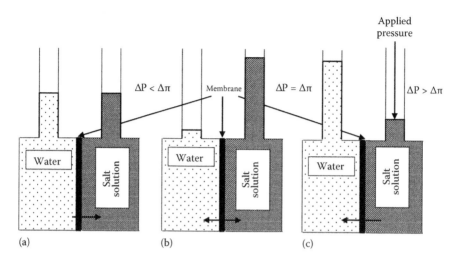

FIGURE 1.3
Diagram of osmosis and reverse osmosis processes. (a) Osmosis. (b) Osmotic equilibrium. (c) Reverse osmosis.

moves from an area of low solute concentration (i.e., pure water having high chemical potential) through a membrane to an area of high solute concentration (low chemical potential). If the level of the salt solution is kept at a certain high position, depending upon the solute concentration, the transport of water through the membrane from the lower concentration to the higher concentration side stops due to osmotic equilibrium (Figure 1.3b). The movement of a pure solvent to equalize solute concentrations on each side of a membrane generates osmotic pressure (Figure 1.3c). Applying an external pressure to reverse the natural flow of pure solvent is thus reverse osmosis.

1.6 Gas Separation

Over the past few decades, gas separation using porous and dense ceramic membranes has drawn a great deal of interest from researchers, especially greenhouse gas separation [32], hydrogen separation [33–35], and oxygen separation [36–39] due to their high resistibility to thermal shock, high mechanical strength, and durability. However, the problem associated with ceramic membrane is the manufacturing cost. Gas mixtures can be separated using either porous or dense ceramic membranes. Dense solid oxide ceramic membranes are the most suitable for oxygen and hydrogen separation and are made of ionic conducting or crystalline materials like modified zirconia (by adding sulfur, tungsten, and molybdate as promoter) and perovskites.

Porous ceramic membranes are prepared by alumina, silica, zirconia, titania, zeolites, etc. Generally, the selection of a ceramic membrane (porous or dense) depends on the operating temperature and driving force; the choice of raw material depends on the manufacturing cost, permeance, and selectivity of the desired product, and on the thermal, mechanical, and chemical sustainability.

A schematic diagram for the gas separation using different ceramic membrane modules (flat, tubular) is shown in Figure 1.4. An elevated pressure is maintained inside the membrane module to pass the feed stream to be separated from one side to the other side of the membrane held at a lower pressure. The driving force for the transport of gas through the membrane is the difference in pressure, which also causes a difference in the relative transport rates of the feed gases. Highly diffusive gas components are kept at a low pressure side (permeate stream), whereas the less diffusive gases are kept at a high pressure side (retentate stream) inside the membrane module.

The mass transport mechanism for separation of gases through ceramic membranes varies and basically depends on membrane structure, the interaction between the membrane and the feed gas, and overall operating conditions such as flow rate of feed components, pressure, temperature, and morphological properties of the membrane (pore size and porosity).

For dense ceramic membranes, a surface reaction and then diffusion in the membrane bulk are the steps to describe the mechanism of transport of gases. For porous ceramic membranes, the mass transfer depends on morphological properties such as pore size and porosity of the membrane and the size of the diffusing molecules. Knudsen diffusion and convective flow are the transport mechanisms for mesoporous and microporous membranes, described by the dusty gas model (DGM) [40]. The transport of gas through a microporous membrane mainly depends on the interaction between the

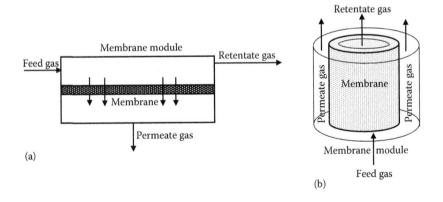

FIGURE 1.4
Schematic representation of (a) flat and (b) tubular ceramic membrane for separation of gas mixture.

diffusing molecules and the pore structure of the membrane, explained by Stefan–Maxwell theory [41].

1.7 Pervaporation

Pervaporation (PV) is a membrane separation process where both selective permeation of one or more species in a volatile liquid mixture and evaporation of its subsequent permeate combine together in a single operation. Firstly, a volatile liquid mixture gets preferentially sorbed on the feed-side membrane. Secondly, the solute diffuses through the membrane and then vaporizes at the product side of the membrane under vacuum. Lastly, the vapor product, which contains the desired compound in a high concentration, transfers into a condenser where it is condensed and recovered as liquid (Figure 1.5).

The overall pervaporation process involves steps in a series [42]:

1. Transport of a species from bulk of the feed solution to the surface of the membrane
2. Selective sorption of the species at the membrane surface
3. Diffusional transport (rate-controlling step) of the species through the membrane

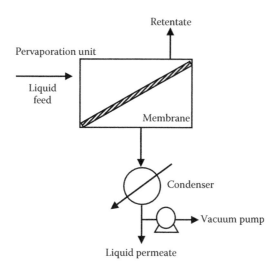

FIGURE 1.5
A diagram of a pervaporation unit.

4. Desorption of the species at the permeate side of the membrane

5. Transport of the species from the membrane surface to the bulk of permeate

Ceramic membranes for the pervaporation process show more advantages than polymeric membranes, such as excellent operating capacity at higher temperatures, no swelling and a more constant performance due to high mechanical stability, and excellent chemical stability, thus operating in acidic and alkaline media, high flux, and high selectivity [43]. The pervaporation process is widely used in separation of azeotropic mixtures, heat-sensitive products, and mixture of components having close boiling points.

1.8 Hybrid Membranes

In a hybrid membrane, a polymer matrix (hydrophobic or hydrophilic)/metal ion or a catalytic layer (macro-, meso-, or microporous) is introduced over a ceramic support to increase the selectivity—yield as well as conversion—to create an interface between two phases to maintain a controlled operation for different applications such as wastewater treatment, control of dissolved gases in liquids, recovery of aroma compounds, removal of acid gases and volatile organic compounds (VOCs), etc. Hybrid membranes are expected to play a key role in the strategy of process intensification, including high efficiency; ease of operation; high selectivity and permeability; easy control and high operational flexibility; low energy consumption; good thermal, chemical, and mechanical sustainability; easy scale-up; etc. Membrane crystallizers, membrane distillation, membrane emulsifiers, membrane extractors, membrane reactors, etc., can be designed and combined with the conventional membrane operations to obtain a hybrid membrane for advanced separation processes.

1.8.1 Membrane Crystallizer

Some critical problems in the design of an existing industrial crystallization process are the strong interaction between hydrodynamics and crystallization kinetics; nonuniform distribution of process parameters, such as temperature, turbulence, and supersaturation in a crystallizer body; and inhomogeneous mixing of slurry. These are the reasons for introducing membrane crystallization (MCr). Membrane crystallization is a novel concept related to the execution of membrane technology in crystallization processes. Some significant advantages make this methodology of increasing interest in chemical as well as biochemical applications [44,45]. This concept offers fast crystallization rates and reduces induction time, well-controlled

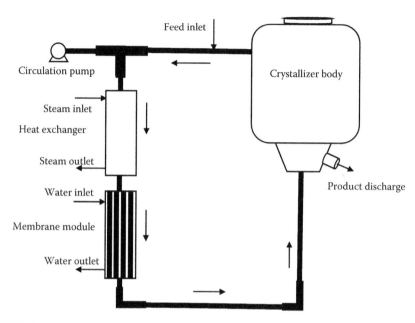

FIGURE 1.6
A unit of membrane crystallizer.

nucleation and growth kinetics, membrane surface supporting heterogeneous nucleation, fine variation of the level and rate of supersaturation,
selection of polymorphic forms, introduction of the MCr concept to antisol-
vent crystallization, and production of cocrystals [46].

In a membrane crystallizer (Figure 1.6), the crystallizing solution is in direct
contact with the membrane matrix, which acts as a selective barrier for solvent evaporation, controlling the rate for the generation of the supersaturation. Therefore, a solute–membrane interaction is likely to occur, depending
on the fluid dynamic regime [47].

1.8.2 Membrane Emulsifier

Emulsions play a significant role for food, cosmetic and pharmaceutical applications. Conventionally, they are prepared using rotor-stator systems, colloid
mills and high pressure homogenizers. The selection is generally said by the
application of the resulting emulsion, the apparent viscosity, the amount of
mechanical energy required and the heat-transfer demands [48].

Membrane emulsification is a comparatively novel technology [49,50]; it
has received increasing attention over the last 10 years, mainly addressed
to producing shear-sensitive monodisperse emulsion droplets with a narrow pore size distribution for pharmaceutical, food, and cosmetic industries [51,52] due to low shear stresses and energy input, as well as simple

and effective design. The unique feature of this process is that the resulting droplet size is controlled mostly by the choice of membrane and not by the generation of turbulent droplet breakup. It is applicable to both oil-in-water (o/w) and water-in-oil (w/o) emulsions.

The membrane emulsification procedure is shown in Figure 1.7. The dispersed phase is forced through the pores of a microporous membrane; however, the continuous phase flows along the membrane surface. Droplets grow at pore openings and then detach until they reach a certain size. This phenomenon is measured by the balance between the driving pressure and the drag force on the droplet from the flowing continuous phase, the buoyancy of the droplet, and the interfacial tension forces [51]. Due to the action of interfacial tension, the droplets at a pore have a tendency to form a spherical shape, but some distortion may possibly take place depending on the flow rate of the continuous phase and the contact angle between the droplet and membrane surface [53]. The final droplet size and size distribution are determined by the pore size and size distribution of the membrane and by the degree of coalescence, both at the membrane surface and in the bulk solution.

A schematic diagram of a typical small-scale membrane emulsification system, consisting of a tubular microfiltration membrane module, a gear pump, a feed vessel, a heat exchanger for recirculation of the continuous phase, and a pressurized (N_2) oil container, is shown in Figure 1.8. The oil phase (to be dispersed) is forced through the pores of the membrane into the aqueous continuous phase by using gas pressure, which circulates through the middle of the membrane.

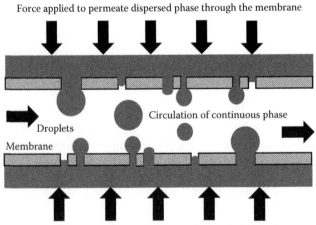

FIGURE 1.7
Membrane emulsification process: a schematic diagram.

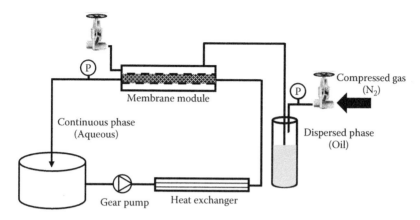

FIGURE 1.8
A schematic diagram of a small-scale membrane emulsification setup.

The influence of different process parameters, such as membrane type, average pore size and porosity, number of active pores, thickness, cross-flow velocity, transmembrane pressure, interfacial tension, shear stress, and emulsifier on the performance of the membrane emulsifier, has been studied. Emulsions with narrow emulsion droplet size distributions with average droplet size ranges between 2 and 10 times the nominal membrane pore diameter have been produced by careful selection of these parameters [54,55].

Traditional emulsification methods, like homogenization and rotor–stator systems, requires high energy to form droplets of a given size compared to the membrane emulsification process and is the significant advantage of the latter process. But one of the main limitations of this process is the low level of dispersed phase flux through the membrane during industrial scale-up, specifically for small submicron droplets [56].

1.8.3 Membrane Reactors

A membrane reactor is a plug-flow reactor that contains an additional cylinder of some porous material within it, similar to the tube within the shell of a shell-and-tube heat exchanger. This porous inner cylinder is the membrane that gives the membrane reactor its name. With respect to conventional reactors, a membrane reactor (MR) permits the improvement of performance in terms of reaction conversion, products' selectivity, and so on. In fact, by means of the so-called "shift effect," the thermodynamic equilibrium restrictions can be overcome. At the least, MRs' behavior could be the same as that of a conventional reactor working at the same MR operating conditions. There are a variety of reactor configurations in the classification of membrane reactors according to the design and application of a membrane reactor to the particular reaction; these are listed in Table 1.2.

TABLE 1.2

Classification of Membrane Reactors

Description	Features
CMR: catalytic membrane reactor	• A membrane with an intrinsically catalytic layer or a membrane prepared by catalytic material • Both separation and reaction occur at the membrane surface
CNMR: catalytic nonpermselective membrane reactor	• A membrane providing catalyst site but that cannot separate every substance • Mostly acts as reactant distributor rather than as separator
PBMR: packed-bed membrane reactor	• Catalyst packed either in the interior or exterior of the membrane volume • Membrane acts as reactant distributor
PBCMR: packed-bed catalytic membrane reactor	• Catalyst packed either in the interior or exterior of the membrane volume • Membrane prepared by catalytic material and functions to separate certain substances
FBMR: fluidized-bed membrane reactor	• Similar to PBMR but catalyst is not packed within the reactor • Has better temperature control than PBMR, especially for exothermic process
FBCMR: fluidized-bed catalytic membrane reactor	• Similar to FBMR but membrane with catalytic properties

Source: J.G. Sanchez Marcano, T.T. Tsotsis. *Catalytic Membranes and Membrane Reactors*, Wiley-VCH Verlag GmbH, Weinham, Germany, 2002.

In a catalytic membrane reactor (CMR), the membrane exhibits catalytic activity, causing the location of reaction and separation to coincide. The catalytic activity of the membrane can be inherent to the membrane material or can be achieved by coating the membrane with a catalytically active material. A catalytic nonpermselective membrane reactor (CNMR) has catalytic properties but only distributes the reactant for the enhancement of yield and selectivity of the product. It is only used to provide a well-defined reactive interface. In the recent past, the most used reactor configuration in research study is the packed-bed membrane reactor (PBMR), in which the membrane provides only the separation function. The membrane might separate the desired product from the mixture products downstream, or simply distribute the reactant for the purpose of increasing reaction site. In order to provide an additional catalytic function in the PBMR configuration, a packed-bed catalytic membrane reactor (PBCMR) is introduced. For better control of process temperature, the packed bed should be replaced by a fluidized bed (FBMR or FBCMR).

Biotechnology is another area in which membrane-based reactive separations are attracting great interest. There, membrane processes are coupled with industrially important biological reactions. These include the broad class of fermentation-type processes, widely used in the biotechnology industry for the production of amino acids, antibiotics, and other fine chemicals. Membrane-based reactive separation processes are of interest for the continuous elimination of metabolites and the immobilization of bacteria, enzymes, or animal cells in the production of many high-value added chemicals.

1.8.3.1 Membrane Bioreactors

Membrane bioreactors (MBRs) have been extensively studied in the food and pharmaceutical industries. The dairy industry, in particular, has been a pioneer in the use of microfiltration, ultrafiltration, nanofiltration, and reverse osmosis membranes.

Applications include the processing of various natural fluids (milk, blood, fruit juices, etc.), the concentration of proteins from milk, and the separation of whey fractions, including lactose, proteins, minerals, and fats. These processes are typically performed at low-temperature and -pressure conditions making use of commercial membranes [57].

MBRs are finding fertile ground for application in biochemical synthesis for the production of a broad spectrum of products. These range from food, liquid fuels (e.g., ethanol), and plant metabolites to fine chemicals, including medical products, flavoring agents, food colors, fragrances, etc. For the synthesis of high-purity pharmaceutical and food products, membranes may offer advantages over the more conventional separation techniques such as distillation, evaporation, crystallization, etc. This is because they are simpler and less energy intensive, and they can operate under the mild conditions required to maintain biological or enzymatic activity for the synthesis of biochemicals, which are sensitive to heat.

The way in which membranes (in various forms, i.e., cylindrical, coaxial, flat sheet, spiral wound, hollow fiber, etc.) couple with the bioreactor depends on the role of the membrane performance. The simplest configuration consists of two separate but coupled units, one being the bioreactor and the other the membrane module. The alternate configuration involves coupling of membrane and bioreactor into the same unit [58].

A growing application of MBRs is in wastewater treatment. Conventionally, wastewater treatment is carried out either by physicochemical techniques or by biological processes. The physicochemical techniques often simply transfer the contaminants in the wastewater streams into a different medium that must, itself, be disposed of or, when it is destroyed, often also generates toxic by-products, which are difficult to eliminate. The biological processes have an advantage, in that they transform the complex organic contaminants into simple, harmless gaseous or water-soluble compounds, together with residual sludge. On the other hand, the conventional biological treatment at the same point has the disadvantage that one must physically separate the biocatalyst from the treated water. However, for heavily polluted wastewaters, fast growing biomass clogs the beds and results in bed shutdowns and the need for frequent regeneration.

Though MBRs offer advantages over the conventional bioreactors, these are not completely free from problems. One such key problem relates to changes in biocatalyst activity. This is a serious concern for whole-cell MBRs when the cells are immobilized in the membrane's pore structure. For enzymatic MBRs, an important problem is the intrinsic decrease in activity as a result of the enzyme's immobilization on the membrane support or the grafting onto various macromolecules in order to increase its molecular size and improve its retention by the membrane. One of the most serious problems encountered with MBRs is biofouling, which typically manifests itself by a dramatic decrease over time in permeate flow. It may be caused by adsorption on the membrane surface and in its internal porous structure of the various metabolites that cells produce, as well as of the coagulated proteins from lysed cells. These accumulate in the reactor over time and also tend to increase the solution viscosity. Another cause of biofouling is pore plugging, when cells are fixed in the membrane—thus considerably decreasing mass transfer [59].

1.8.3.2 Catalytic Membrane Reactors

A large amount of research work in the field of membrane reactors has been published since 1960, when MRs were introduced only as a concept in the field of membrane technology. In such an integrated process, the membrane was used as an active participant in a chemical transformation for increasing the reaction rate, selectivity, and yield. Thus, membrane-based reactive separation processes have been introduced—that is, a combination of both separation and reaction in a single unit, generally termed a catalytic membrane reactor.

The significant benefit of using CMRs is the capacity to maintain the synergistic effect between reaction and separation [60]. Two important parameters are always in consideration at the time of preparation of a catalytic membrane: selectivity and permeability. Selectivity means the ability of a membrane to allow only some, but not other, substances to enter the cell of the membrane, which is related to purity of the substance that is being separated. Permeability is the rate of diffusion of certain molecules or ions passed through the membrane per unit area per unit time.

The conventional membrane reactor systems used before 1980s are shown in Figure 1.9. Figure 1.10 displays the new concept of having reaction and separation together in a single unit. Other advantages of catalytic membrane reactors are less energy consumption, simple design, and ability of achieving high yield or selectivity—all of which make this concept popular [57].

Several investigations have discussed the concept of CMRs and their various potential benefits, such as increased reaction rate; enhanced selectivity and yield for a variety of reactions involving basic functions of

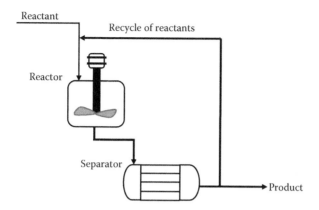

FIGURE 1.9
Conventional membrane reactor system.

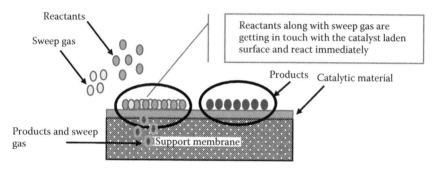

FIGURE 1.10
Schematic of an integrated membrane reactor system that combines reactor and separator into a single unit.

membrane as extractor, distributor, or contactor [4]; and different configurations [61,62]. Numerous review articles related to the applications of CMRs have been published in the recent past [63]. But there is much scope for further improvement of CMRs in different applications, especially in petroleum industries.

In the petrochemical industry, olefins such as ethylene and propylene are the most important chemicals used for the production of polyolefin—namely, polyethylene, polypropylene, styrene, ethyl benzene, ethylene dichloride, acrylonitrile, and isopropanol. An important step in the manufacture of olefins is large-scale separation of the olefin from the corresponding paraffin [64].

Furthermore, dehydrogenation, oxidative coupling of methane, steam reforming of methane, removal of sulfur (the Claus reaction), and the water gas shift reaction are some important reactions in the petrochemical industry.

Membrane gas separation is attractive because of its simplicity and low energy cost, but it has one major drawback: the reverse relationship between selectivity and permeability. Petrochemical waste streams may contain phenolic compounds or aromatic amines. These are highly toxic and, at high concentrations, inhibit biological treatment. The membrane aromatic recovery system (MARS) is a relatively new process for recovery of aromatic acids and bases. Wastewater in the petrochemical industry is currently treated by an activated sludge process with pretreatment of oil/water separation [57]. Tightening effluent regulations and increasing need for reuse of treated water have generated interest in the treatment of petrochemical wastewater with the advanced membrane bioreactor process.

1.9 Flow Patterns

There are two types of flow patterns and filtration arrangements in UF and MF membranes: (a) dead-end filtration and (b) cross-flow filtration, as shown in Figure 1.11.

In a dead-end filtration system, the feed flow is parallel to the membrane surface. The solute or the rejected particles remain on the membrane, forming a gel or cake layer. This phenomenon enhances the accumulation of particles on the membrane surface and causes increase in membrane resistance. The resistance can only be minimized by applying pressure to maintain the flow.

In the cross-flow filtration arrangement, feed flows in the parallel direction to the membrane surface. Most of the retained particles or the solutes are swept away with the flowing feed side liquid. Recirculation of the retentate to enhance the concentration can be achieved in this arrangement. This flow pattern involves less or negligible accumulation of solid particles on the membrane and is favored.

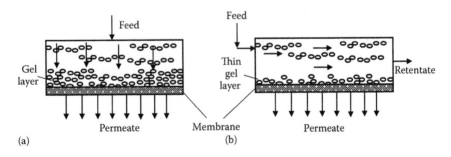

FIGURE 1.11
(a) Dead-end filtration arrangement and (b) cross-flow system.

1.10 Advantages and Disadvantages of Ceramic Membranes

Compared to polymeric membranes, ceramic membranes have widespread use in the field of chemical engineering and biotechnology due to the following properties:

- Chemical, mechanical, and thermal stability
- Ability of steam sterilization and back flushing
- High abrasion resistance
- High fluxes
- High durability
- Bacteria resistance
- Possibility of regeneration

Disadvantages of ceramic membranes are their heavy weight and the considerable production costs of ceramic raw materials, which causes increase in the membrane fabrication cost. However, nowadays, cheap raw materials are used to solve this problem.

1.11 Summary

This chapter provided an overview of ceramic membranes that included classifications of traditional ceramic membranes and their applications and advancement of ceramic membranes in terms of configurations and novel applications. Operating modes of membranes has been summarized precisely in this segment.

- Introduction to ceramic membranes: definition, types, and structures of ceramic membranes.
- Basic concepts, structural geometry (porous or dense), material of construction and the applications of conventional membrane separation techniques such as microfiltration, ultrafiltration, nanofiltration, reverse osmosis, gas separation, and pervaporation.
- Advanced ceramic membranes such as membrane crystallizers and emulsifiers have a process and advantages in separation over conventional separation methods.
- Definition, configuration, classifications, and applications of membrane reactors were given.

- Catalytic membrane reactor: This is a membrane with a catalytic layer naturally or a membrane prepared by catalytic material; both separation and reaction occur simultaneously at the membrane surface.
- Catalytic nonpermselective membrane reactor: This is a membrane with a catalyst site that cannot separate every substance; it mostly acts as a reactant distributor rather than as a separator.
- Packed-bed membrane reactor: This catalyst is packed either in the interior or exterior of the membrane surface. A PBMR acts as a reactant distributor like a CNMR.
- Packed-bed catalytic membrane reactor: This catalyst is packed either in the interior or exterior of the membrane volume. The membrane is prepared by catalytic materials and functions to separate certain substances.
- Fluidized-bed membrane reactor: This is similar to a PBMR but the catalyst is not packed within the reactor. It has better temperature control than a PBMR, especially for exothermic processes.
- Fluidized-bed catalytic membrane reactor: This is similar to an FBMR but the membrane has catalytic properties.
- The different flow patterns are dead end and cross flow.
- Merits and demerits of ceramic membranes were discussed.

References

1. *Encyclopaedia Britannica Online*, "whiteware pottery" https://www.britannica.com/art/whiteware, retrieved June 30, 2015.
2. G. Geiger, Introduction to ceramics. *American Ceramic Society Bulletin*, 69 (7) (1990) 1131.
3. K. Li, *Ceramic membranes for separation and reaction*, John Wiley & Sons Ltd., Chichester, England, 2007, ISBN 978-0-470-01440-0.
4. H.W. Liepman, Gas kinetics gas dynamics of orifice flow. *Journal of Fluid Mechanics*, 10 (1961) 65–79.
5. J. Kong, K. Li, An improved gas permeation method for characterizing and predicting the performance of microporous asymmetric hollow fibre membranes used in gas absorption. *Journal of Membrane Science*, 182 (1–2) (2001) 271–281.
6. M. Knudsen, Die gesetze der molekularstromung und der inner reibunsstromung der gase durch rohren. *Annals of Physics*, 28 (1909) 75–130.
7. R.D. Present, *Kinetic theory of gases*, McGraw–Hill, New York, 1958.
8. C.R. Metz, *Physical chemistry*, McGraw–Hill, New York, 1976.
9. B.K. Nandi, B. Das, R. Uppaluri, M.K. Purkait, Microfiltration of mosambi juice using low cost ceramic membrane. *Journal of Food Engineering*, 95 (4) (2009) 597–605.

10. B.K. Nandi, B. Das, R. Uppaluri, Clarification of orange juice using ceramic membrane and evaluation of fouling mechanism. *Journal of Food Process Engineering*, 35 (3) (2012) 403–423.

11. B.K. Nandi, R. Uppaluri, M.K. Purkait, Treatment of oily waste water using low-cost ceramic membrane: Flux decline mechanism and economic feasibility. *Separation Science and Technology*, 44 (12) (2009) 2840–2869.

12. B.K. Nandi, A. Goswami, M.K. Purkait, Adsorption characteristics of brilliant green dye on kaolin. *Journal of Hazardous Materials*, 161 (1) (2009) 387–395.

13. V. Singh, M.K. Purkait, V.K. Chandaliya, P.P. Biswas, P.K. Banerjee, C. Das, Development of membrane based technology for the separation of coal from organic solvent. *Desalination*, 299 (2012) 123–128.

14. H.P. Hsieh, *Inorganic membranes for separation and reaction*. Membrane science and technology series, vol. 3, Elsevier, Amsterdam, the Netherlands, 1996, ISBN: 978-0-444-81677-1.

15. N.F.D. Aba, J.Y. Chong, B. Wang, C. Mattevi, K. Li, Graphene oxide membranes on ceramic hollow fibers—Microstructural stability and nanofiltration performance. *Journal of Membrane Science*, 484 (2015) 87–94.

16. X. Da, X. Chen, B. Sun, J. Wen, M. Qiu, Y. Fan, Preparation of zirconia nanofiltration membranes through an aqueous sol-gel process modified by glycerol for the treatment of wastewater with high salinity. *Journal of Membrane Science*, (2016) doi:10.1016/j.memsci.2015.12.068.

17. S. Benfer, U. Popp, H. Ritcher, C. Siewert, G. Tomandl, Development and characterization of ceramic nanofiltration membranes. *Separation and Purification Technology*, 22–23 (1–3) (2001) 231–237.

18. R. Weber, H. Chmiel, V. Mavrov, Characteristics and application of new ceramic nanofiltration membranes. *Desalination*, 157 (1–3) (2003) 113–125.

19. G. Zhu, Q. Jiang, H. Qi, N. Xu, Effect of sol size on nanofiltration performance of a sol–gel derived microporous zirconia membrane. *Chinese Journal of Chemical Engineering*, 23 (1) (2015) 31–41.

20. W. Puthai, M. Kanezashi, H. Nagasawa, K. Wakamura, H. Ohnishi, T. Tsuru, Effect of firing temperature on the water permeability of SiO_2–ZrO_2 membranes for nanofiltration. *Journal of Membrane Science*, 497 (2016) 348–356.

21. T. Tsuru, T. Sudoh, T. Yoshioka, M. Asaeda, Nanofiltration in non-aqueous solutions by porous silica–zirconia membranes. *Journal of Membrane Science*, 185 (2) (2001) 253–261.

22. T. Tsuru, M. Narita, R. Shinagawa, T. Yoshioka, Nanoporous titania membranes for permeation and filtration of organic solutions. *Desalination*, 233 (1–3) (2008) 1–9.

23. A. Alem, H. Sarpoolaky, M. Keshmiri, Sol-gel preparation of titania multilayer membrane for photocatalytic applications. *Ceramics International*, 35 (5) (2009) 1837–1843.

24. J. Sekulic, J.E. ten Elshof, D.H.A. Blank, A microporous titania membrane for nanofiltration and pervaporation. *Advanced Materials*, 16 (17) (2004) 1546–1550.

25. I. Voigt, G. Fischer, P. Puhlfurss, M. Schleifenheimer, M. Stahn, TiO_2–NF membranes on capillary supports. *Separation and Purification Technology*, 32 (1–3) (2003) 87–91.

26. X. Chen, W. Zhang, Y. Lin, Y. Cai, M. Qiu, Y. Fan, Preparation of high-flux γ-alumina nanofiltration membranes by using a modified sol-gel method. *Microporous and Mesoporous Materials*, 214 (2015) 195–203.

27. S. Condom, A. Larbot, S. Alami Younssi, M. Persin, Use of ultra- and nanofiltration ceramic membranes for desalination. *Desalination,* 168 (2004), 207–213.
28. T. Van Gestel, B. Van der Bruggen, A. Buekenhoudt, C. Dotrement, J. Luyten, C. Vandecasteele, G. Maes, Surface modification of γ-Al_2O_3/TiO_2 membranes. *Journal of Membrane Science,* 224 (1–2) (2003) 3–10.
29. S. Alamiyounssi, A. Larbot, M. Persin, J. Sarrazin, L. Cot, Rejection of mineral salts on gamma-alumina nanofiltration membrane application to environmental process. *Journal of Membrane Science,* 102 (1995) 123–129.
30. Y. Cai, X. Chen, Y. Wang, M. Qiu, Y. Fan, Fabrication of palladium–titania nanofiltration membranes via a colloidal sol-gel process. *Microporous and Mesoporous Materials,* 201 (2015) 202–209.
31. H. Qi, S. Niu, X. Jiang, N. Xu, Enhanced performance of a macroporous ceramic support for nanofiltration by using α-Al_2O_3 with narrow size distribution. *Ceramics International,* 39 (3) (2013) 2463–2471.
32. S. Lee, J.-W. Choi, S.-H. Lee, Separation of greenhouse gases (SF_6, CF_4 and CO_2) in an industrial flue gas using pilot-scale membrane. *Separation and Purification Technology,* 148 (2015) 15–24.
33. Z. Zhu, J. Hou, W. He, W. Liu, High-performance $Ba(Zr_{0.1}\,Ce_{0.7}Y_{0.2})O_{3-\delta}$ asymmetrical ceramic membrane with external short circuit for hydrogen separation. *Journal of Alloys and Compounds,* 660 (2016) 231–234.
34. D. van Holt, E. Forster, M.E. Ivanova, W.A. Meulenberg, M. Müller, S. Baumann, R. Vaßen, Ceramic materials for H_2 transport membranes applicable for gas separation under coal-gasification-related conditions. *Journal of the European Ceramic Society,* 34 (10) (2014) 2381–2389.
35. W.A. Rosensteel, S. Ricote, N.P. Sullivan, Hydrogen permeation through dense $BaCe_{0.8}Y_{0.2}O_{3-\delta}$–$Ce_{0.8}Y_{0.2}O_{2-\delta}$ composite-ceramic hydrogen separation membranes. *International Journal of Hydrogen Energy,* (2016) doi:10.1016/j.ijhydene.2015.11.053.
36. K.J. Yoon, O.A. Marina, Highly stable dual-phase $Y_{0.8}Ca_{0.2}Cr_{0.8}Co_{0.2}O_3$–$Sm_{0.2}Ce_{0.8}O_{1.9}$ ceramic composite membrane for oxygen separation. *Journal of Membrane Science,* 499 (2016) 301–306.
37. N. Kosinov, J. Gascon, F. Kapteijn, E.J.M. Hensen, Recent developments in zeolite membranes for gas separation. *Journal of Membrane Science,* 499 (2016) 65–79.
38. C. Wu, Y. Gai, J. Zhou, X. Tang, Y. Zhang, W. Ding, C. Sun, Structural stability and oxygen permeability of $BaCo_{1-x}Nb_xO_{3-\delta}$ ceramic membranes for air separation. *Journal of Alloys and Compounds,* 638 (2015) 38–43.
39. F. Yang, H. Zhao, J. Yang, M. Fang, Y. Lu, Z. Du, K. Świerczek, K. Zheng, Structure and oxygen permeability of $BaCo_{0.7}Fe_{0.3-x}In_xO_{3-\delta}$ ceramic membranes. *Journal of Membrane Science,* 492 (2015) 559–567.
40. E.A. Mason, A.P. Malinauskas, *Gas transport in porous media: The dusty gas model,* Elsevier, Amsterdam, the Netherlands, 1983.
41. R. Taylor, R. Krishna, *Multicomponent mass transfer,* Wiley series in chemical engineering, New York, 1993.
42. B.G. Gonzalez, I.O. Utribe, Mathematical modeling of the pervaporative separation of methanol–methyl–terbutyl ether mixtures. *Industrial Engineering and Chemistry Research,* 40 (2001) 1720–1731.
43. H.M. van Veen, Y.C. van Delft, C.W.R. Engelen, P.P.A.C. Pex, Dewatering of organics by pervaporation with silica membranes. *Separation and Purification Technology,* 22–23 (2001) 361–366.

44. E. Drioli, E. Curcio, A. Criscuoli, G. Di Profio, Integrated system for recovery of $CaCO_3$, NaCl and $MgSO_4.7H_2O$ from nanofiltration retentate. *Journal of Membrane Science*, 239 (2004) 27–38.

45. G. Di Profio, E. Curcio, A. Cassetta, D. Lamba, E. Drioli, Membrane crystallization of lysozyme: Kinetic aspects. *Journal of Crystal Growth*, 257 (2003) 359–369.

46. E. Drioli, G. Di Profio, E. Curcio, Progress in membrane crystallization. *Current Opinion in Chemical Engineering*, 1 (2) (2012) 178–182.

47. E. Curcio, E. Fontanova, G. Di Profio, E. Drioli, Influence of the structural properties of polyvinylidenefluoride (PVDF) membranes on the heterogeneous nucleation rate of protein crystals. *Journal of Physical Chemistry B*, 110 (2006) 12438–12445.

48. M.J. Lynch, W.C. Griffin, Food emulsions, in: *Emulsions and emulsion technology*. Surfactant science series, vol. 6(1), Marcel Dekker, New York, 1974, Chapter 5, pp. 249–289.

49. T. Nakashima, M. Shimizu, Advanced inorganic separative membranes and their developments. *Chemical Engineering Symposium Series*, 21 (1988) 93–99.

50. T. Nakashima, M. Shimizu, M. Kukizaki, *Membrane emulsification operational manual*, 1st ed., Industrial Research Institute of Miyazaki Prefecture, Miyazaki, 1991.

51. V. Schroder, O. Behrend, H. Schubert, Effect of dynamic interfacial tension on the emulsification process using microporous ceramic membranes. *Journal of Colloid and Interface Science*, 202 (1998) 334–340.

52. S.M. Joscelyne, G. Trägårdh, Food emulsions using membrane emulsification: Conditions for producing small droplets. *Journal of Food Engineering*, 39 (1999) 59–64.

53. S.J. Peng, R.A. Williams, Controlled production of emulsions using a cross-flow membrane. *Particle & Particle Systems Characterization*, 15 (1998) 21–25.

54. E. Drioli, A. Criscuoli, E. Curcio, *Membrane contactors: Fundamentals, applications and potentialities*. Membrane science and technology series 11, Elsevier B.V., Amsterdam, the Netherlands, 2006.

55. S.M. Joscelyne, G. Trägårdh, Membrane emulsification—A literature review. *Journal of Membrane Science*, 169 (1) (2000) 107–117.

56. P. Walstra, P.E.A. Smulders, Emulsion formation, in: B.P. Binks (ed.), *Modern aspects of emulsion science*, Royal Society of Chemistry, Cambridge, UK, 1998.

57. J.G. Sanchez Marcano, T.T. Tsotsis, *Catalytic membranes and membrane reactors*, Wiley-VCH Verlag, GmbH, Weinham, Germany, 2002.

58. J.M. Engasser, Reacteurs a enzymes et cellules inmobilisees, in *Biotechnologie, technique et documentation*, Lavoisier, Paris, Chapter 4.2, 1988.

59. J.H. Hanemaaijer, J. Stahouders, S. Vissar, in *Proceedings of 4th European Congress*, in: O.M. Neijssel, R.R. van der Meer, K.Ch.M. Luyben (eds.), *Biotechnology*, vol. 1, p. 119, Elsevier, Amsterdam, the Netherlands, 1987.

60. Market report, global membrane technology market, Acmite Market Intelligence, ca. 550 pp. (2013).

61. A. Julbe, D. Farrusseng, C. Guizard, Porous ceramic membranes for catalytic reactors—Overview and new ideas. *Journal of Membrane Science*, 181 (2001) 3–20.

62. T. Westermann, T. Melin, Flow-through catalytic membrane reactors—Principles and applications. *Chemical Engineering Process*, 48 (2009) 17–28.

63. A.G. Dixon, Innovations in catalytic inorganic membrane reactors, in: J.J. Spivey (ed.), *Catalysis*, Chapter 2, pp. 40–92, vol. 14, Specialist Periodical Reports, RSC Publishing (1999).
64. I. Pinnau, L.G. Toy, Solid polymer electrolyte composite membranes for olefin—Paraffin separation. *Journal of Membrane Science*, 184 (2001) 39–48.

2

Fabrication of Ceramic Membranes

2.1 Introduction to Low-Cost Ceramic Membranes with Examples

In this chapter, several fabrication processes of ceramic membranes, use of raw materials, pore-formers, binders, and additives and their effects on the ceramic membrane fabrication are exemplified. Different shapes and modules of membranes are also discussed. Manufacturing cost of ceramic membranes is also mentioned in this chapter.

The cost of the precursors, such as α-alumina, γ-alumina, zirconia, titania, and silica usually used in ceramic processing is significantly high and therefore contributes to the operating cost of membrane modules for industrial purposes. To overcome the issue of membrane cost, research in the recent past for the fabrication of ceramic membranes has been focused toward the utilization of cheaper raw materials, such as natural raw clay [1,2], fly ash [3], cordierite powder [3,4], apatite powder [5], dolomite [6,7], calcite [8], and kaolin [9–12].

The most popular example of use of ceramic membranes is wastewater treatment for production of drinking water. Water is a major requirement on the earth because, without water, life is impossible. About one billion people are without safe drinking water worldwide. The majority of these people are found in sub-Saharan Africa, South Asia, and East Asia. Numerous lives are lost annually due to consumption of contaminated water. Globally, four billion cases of diarrhea are reported every year, causing 1.8 million deaths, out of which about 90% are children under age 5. The most common source of drinking water for rural people is groundwater from boreholes (deep wells), shallow wells, and springs. Groundwater is usually consumed without any form of treatment. Water is a medium for thousands of microorganisms, some of which cause disease. Pathogens (e.g., bacteria, viruses, protozoa, and helminths) in water cause a variety of diarrhea-related diseases, such as cholera. These pathogens are commonly derived from human fecal material. Approximately 2.2 billion people are without adequate sanitation in the world [13]. In Ethiopia, the majority of people in rural areas and high-density townships in urban areas use pit latrines which are often in a state of disrepair and are unhygienic. In the rainy season, fecal matter from pit latrines and open sources is washed into water bodies, thereby contaminating the water. In urban areas, sanitation facilities fill up and overflow if they are not properly managed. Microbiological water quality from shallow wells (with depths not exceeding 20 m) has been found to be more inferior in the wet season compared to the dry season [14].

Conventional treatment of raw water for the municipal supply of drinking water may include chemical addition, coagulation, flocculation, sedimentation, filtration, and disinfection, usually with chlorine.

The motive for a greater use of membrane filtration systems in this field, such as microfiltration and ultrafiltration, is mostly due to the ability of the membrane to remove pathogenic microorganisms as well as to control the disinfection by-products' (DBPs) precursor.

Bottino et al. [15] have studied the use of microfiltration ceramic membranes for the treatment of raw water drawn from a lake located near Genova. They have investigated the behavior of permeate flux as a function of significant operating variables like temperature, transmembrane pressure, test duration, and membrane retention toward particles, microorganisms, algae, and DBP precursors. Their study proved that the membrane filtration of lake water using porous ceramic membrane is very useful for drinking water production. Suspended solids have completely been removed along with microorganisms (fecal coliforms, total coliforms, and spiked *Escherichia coli*) and algae (*Asterionella, Ceratium, Melosira, Navicula,* etc.), and retention rates of 64% and ca. 56% have been achieved for total organic carbon (TOC) and chloroform, respectively. Retention rates of 99.6% and 56%–100% have also been achieved for turbidity and DBPs, respectively.

Therefore, a substantial improvement of the quality of water to be delivered to the consumer's tap was obtained, and this led to a noticeable reduction of the chlorine demand needed to render the transport and the distribution network of the water hygienically safe [15].

In another study, a ceramic filter was made of a mixture of 77.5% natural clay, 20% fly ash, and 2.5% iron powder and used to treat water contaminated with heavy metals such as iron (Fe) and zinc (Zn). The variables studied were feed flow rate and operation time. The attained reduction percentages of Fe and Zn ions in water samples were more than 99% and 96%, respectively [16].

Saffaj et al. have prepared a low-cost ceramic support membrane with Moroccan natural clay for the removal of salts and dyes (toxic elements) from wastewater to make it drinkable. They have provided optimal inorganic formulations (based on a wet basis) using clay (81.7% w/w), Amidon (10% w/w), Methocel (4% w/w), Amijel (4% w/w), PEG1500 (0.3% w/w), water, and Zusoplast 126/3 for fabricating ceramic supports capable of removal of salts and ionic dyes. The sintering temperature of this work is about 1250°C and the mechanical strength is reported as 10 MPa [1]. In another approach, they used the same composition for ceramic membrane fabrication for economic treatment of wastewater containing toxic compounds [2].

In the area of wastewater treatment, membrane processes are often used in combination with other processes to treat very complex effluents, which often have an important load of organic substances and salt. The membrane process would enhance the water-treated quality in order to reuse water. Khemakhem et al. carried out an experiment in order to reduce the pollution load of the cuttlefish effluent generated from a sea-product freezing factory located in Tunisia that consumes a great amount of water for the washing baths (about 150–200 m^3/day), which is generally discharged in the littoral waters. They have reported optimal inorganic formulations (based on a wet basis) using clay (59% w/w), Methocel (4% w/w), Amijel (4% w/w), cornstarch (8% w/w), and water (25% w/w) for the preparation of ceramic support and utilized for cuttlefish effluent treatment. The sintering temperature of this work has been reported at about 1190°C, with average pore diameters and porosity of about 9.2 μm and 49%, respectively [17].

2.2 Factors: Membrane Fabrication

Fabrication of ceramic membranes depends on several factors, such as raw materials, membrane shapes, and modules. The performance of a membrane (permeability and selectivity) is typically ruled by the pore size, thickness, and surface porosity of the membrane (i.e., the morphology of the membrane). Surface modification of a ceramic membrane is influenced by the fabrication and coating technique as well as the physical and chemical properties of raw materials. The significant factor in ceramic fabrication is the manufacturing cost of the membrane, which is mainly controlled by the cost of raw materials; another factor is energy consumption during the sintering process. A ceramic membrane can be achieved via multiple sintering steps to obtain excellent mechanical strength. It also requires high energy consumption, making ceramic membrane fabrication extremely expensive.

2.2.1 Raw Materials

The subject of ceramic membranes covers extensive ranges of materials. The ancient ceramic raw material was undoubtedly clay. Clay, a sticky and tenacious constituent of soil, forms a clear paste but is not transparent when mixed with water. It can readily be molded into different shapes; if dried, it becomes hard and brittle but can retain its shape. Clay is also available as rock or slate, which are so hard that water penetrates very slowly inside the pores. Now, raw materials for ceramic membrane fabrication are nonmetallic and inorganic compounds, categorized into two parts: oxide and nonoxide. Examples of oxides are silicates such as kaolinite ($Al_2Si_2O_5(OH)_4$) and mullite ($Al_6Si_2O_{13}$) [18–20]. Simple oxides such as zirconia (ZrO_2) [21–23], alumina

(Al_2O_3) [24–26], and titania (TiO_2) [27–29] are used as raw materials for the fabrication of ceramic membrane. The nonoxides are silicon carbide (SiC) and silicon nitride (Si_3N_4) and are receiving attention in fabricating ceramic support membranes due to their ability to bear high mechanical stresses and high temperatures, as well as corrosive environments, suggesting their suitability as support structures for membrane applications [30–32].

2.2.2 Membrane Shapes

The performance of a ceramic membrane always depends on the shape of the membrane. Membrane shape is categorized on the basis of its geometry, such as circular, tubular, cylindrical, rectangular, hexagonal, and wheel-like (cylindrical pipe with a sign of a cross inside) shapes. Here, we will discuss the common shapes of rectangular, circular, cylindrical, or tubular.

2.2.3 Membrane Modules

For different practical applications, ceramic membranes are installed in a suitable device, generally called a membrane module. Basically, membrane modules protect membranes from mechanical damage and provide tightness between streams of permeate and retentate in the high ratio of membrane surface to module volume. According to the membrane shape, various types of membrane modules are available. The choice of membrane modules is based on the following considerations:

- Nature of separation problem
- Ease of cleaning
- Ease of maintenance
- Ease of operations
- Compactness of the system
- Scale-up
- Possibility of membrane replacement

Mainly, four different designs of membrane modules are most commonly used in the field of ceramic membranes:

- Flat sheet
- Cylindrical or tubular
- Hollow fiber
- Monolithic

The selection, design, and operation of membrane modules also include cost of supporting materials and power consumption in pumping.

2.2.4 Fabrication and Coating Techniques

Generally, manufacturing of ceramic membranes involves several steps:

1. Formation of particles by mixing raw materials, either in dry phase or in the form of suspension
2. Packing of particles to obtain a suitable shape
3. Consolidation (hardness or strength) by the sintering process (heat treatment at high temperatures)

A universal procedure for preparation of ceramic membranes as well as preparation of composite membranes by coating a multilayer over a membrane support is shown in Figure 2.1, where some conventional approaches for the fabrication of ceramic membrane are listed, such as extrusion, slip casting, tape casting, and pressing. No matter which fabrication process is selected, the most significant step in preparation of the ceramic membranes is the sintering process to obtain a ceramic membrane or membrane support.

2.2.5 Manufacturing Cost

Ceramic microfiltration/ultrafiltration membranes have been used in various applications, including pharmaceuticals, chemicals, and foods and beverages for 30 years. But the high manufacturing cost is a huge drawback for these membranes and has limited their use to the most challenging applications where polymeric membranes are known to fail. In recent years, low-cost ceramic membranes have been used widely in various applications, such

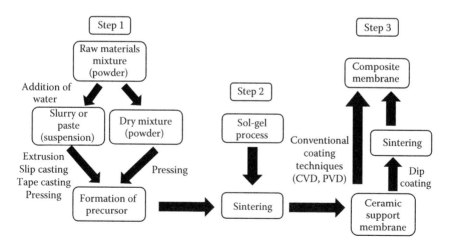

FIGURE 2.1
Flowchart for the preparation of ceramic membrane through three different steps using traditional methods.

as removal of pollutants, separation of oil–water emulsion, oxygen transport, etc. [10–12,33,34], but till now these membranes have not been implemented to modify existing separation processes.

2.3 Ceramic Forming

Forming ceramics depends upon raw materials, including pore-formers and different additives—namely, binders, plasticizers, solvents, dispersants, etc. The desired objective of the forming steps is the production of a ceramic membrane with good shape and size, including uniform particle packing, high strength with flexibility, and an excellent morphology (pore size and porosity).

2.3.1 Use of Major Chemicals

All ceramics need one or more powdered base materials such as kaolin (inexpensive), feldspar ($KAlSi_3O_8–NaAlSi_3O_8–CaAl_2Si_2O_8$) (inexpensive), quartz (expensive), and natural corundum (expensive), along with binders and additives to form into shapes. These base materials provide different properties to the ceramic body. Kaolin provides low plasticity and high refractory properties to the membrane support when it is heated above 450°C to form metakaolinite ($Al_2Si_2O_7$). Mechanical and thermal stability of a membrane can be enhanced by the addition of quartz. Feldspar acts as a flux (a substance added to material to enable it to fuse more readily) and offers a glassy matrix in the early stage of the firing process. It can also improve product hardness, durability, and resistance to chemical corrosion. Natural corundum gives high fusion resistance capacity to ceramics (e.g., high-alumina refractory bricks). Objects of sintered natural corundum are hard, not freely attacked by acids or alkalis at high temperatures, and can resist substantial changes of temperature without cracking. It also has very high electrical resistance, so insulators are made from it [35].

2.3.2 Use of Pore-Formers

Performance of a ceramic membrane is governed by its morphology, which depends on the physical, chemical, and thermal properties of a pore-former. Porous texture in the ceramic is controlled by pore-formers that, under sintering conditions, release carbon dioxide (CO_2) gas. The path taken by the released CO_2 gas thereby creates the porous texture of the inorganic membrane and contributes to the membrane porosity. Calcium carbonate and sodium carbonate are the conventional inorganic pore-formers that fabricate porous ceramic membranes [36–38]. Sawdust—a lignocellulosic, inexpensive material—is used as a pore-former with a cellular microstructure of high porosity, good mechanical strength, stiffness, and toughness. The advantage of using sawdust over conventional pore-formers is the achievement of high porous structure economically [10–12]. Recently, starch has also been used as a pore-former for fabricating microfiltration ceramic membrane application [39].

2.3.3 Use of Additives

Both organic and inorganic additives play important roles in ceramic forming. Organic additives can be synthetic or natural and are used widely in advanced ceramics as these can be decomposed completely from the green body, thus reducing degradation of the final product's microstructure. Inorganic additives are mainly used in the manufacturing of traditional ceramics and these cannot be removed completely after the sintering step. On the basis of specified functions, additives are categorized as solvents (liquids), dispersants, binders, plasticizers, lubricants, and wetting agents.

Solvents are mainly used to provide fluidity of the powder during mixing and forming, and they can be used as dissolving agents for the powders. Solvents are classified as aqueous (water) and organic. Aqueous solvents can be divided into three groups in terms of polarity of molecules: (1) nonpolar, (2) polar, and (3) hydrogen bonding. Nonpolar solvents have no specific orientation of molecules due to minimum intermolecular forces of attraction between atoms of different nonpolar groups. Polar solvents interact more strongly with a polar surface due to the strong electric field between any bond-connecting atoms. Hydrogen bonding forms when a hydrogen atom in a polar bond approaches an atom with an unshared pair of electrons. Organic solvents are selected as a better choice compared to water due to high vapor pressure, low latent heat of vaporization, low boiling point, and low surface tension. But in terms of cost, the aqueous solvents (water) are preferable to the organic solvents. The selection of solvents depends upon various properties—namely, the ability to dissolve other additives, to wet powder, viscosity, evaporation rate, reactivity toward the powder, safety, and cost [40].

Dispersants are used to stabilize suspensions of solid particles in liquid systems against flocculation and are sometimes considered to be defloculants. Dispersants can also enhance the particle concentration for some usable viscosity of the slurry. Based on chemical structure, dispersants can be classified into three classes: (1) inorganic acid salts (e.g., sodium hexametaphosphate, $Na_6P_6O_{18}$, tetrasodium pyrophosphate, $Na_4P_2O_7$, sodium silicate and sodium tetraborate, $Na_2B_4O_7$, etc.), (2) surfactants (oleic acid, stearic acid, sodium oleate, etc.), and (3) low to medium molecular weight polymers (polyethylene oxide [PEO], polyethylene glycol [PEG], polyvinyl alcohol [PVA], etc.).

Binders are employed in the raw material mixture during formation of ceramic to provide strength to the ceramic body by forming bridges among the particles. Binders can provide plasticity to the feed material for supporting the ceramic manufacturing process. Binders can be organic (PVA, PEG, PEO) or inorganic (sodium metasilicate, boric acid). The choice of a binder for a particular forming procedure includes the consideration of several factors, such as molecular weight, binder burnout characteristics, glass transition temperature, compatibility with the dispersant, influence of solvent on the viscosity, solubility in the solvent, and cost.

Plasticizers are lower molecular weight organic substances used to soften the binder in the dry state, thus increasing the flexibility of the ceramic body. In the dry phase, plasticizer and the binder are mixed homogeneously as a single element. In the wet phase (slurry), both plasticizer and the binder must be soluble in the same liquid, forming a homogeneously mixed slurry to provide a better shape to the ceramic. Plasticizer enhances the softening of the binder but reduces the strength of the ceramic, as the plasticizer molecules grow between the polymer chains of the binder, distracting the chain alignment and reducing the van der Waals bonding between adjacent chains.

Lubricants (steric acid, stearates) are used to reduce the friction between the particles themselves or between the particles and die walls in die compaction, extrusion, and injection molding.

A liquid wetting agent can be added with the raw materials in ceramic fabrication to reduce the surface tension of the liquid by improving the wetting of the particles.

2.4 Membrane Modules for Ceramics

Membranes are housed in a suitable device made of either stainless steel or mild steel for practical applications and considered as a membrane module. The ceramic membranes can be cast as tubular, cylindrical, rectangular, circular, or monolithic. For accommodating such kinds of shapes and structures for several applications, different types of membrane modules are available. The most commonly used designs for membrane modules are cylindrical, flat sheet, tubular, and monolithic. Different membrane modules are selected, designed, and fabricated on the basis of design parameters depending upon particular operation, cost of supporting materials, energy consumption, handling simplicity, and replaceability.

2.4.1 Tubular

The tubular membrane module (Figure 2.2a) uses a housing that includes at least one tube containing a semipermeable membrane of tubular configuration supported on the inner surface of a porous pipe such as a glass

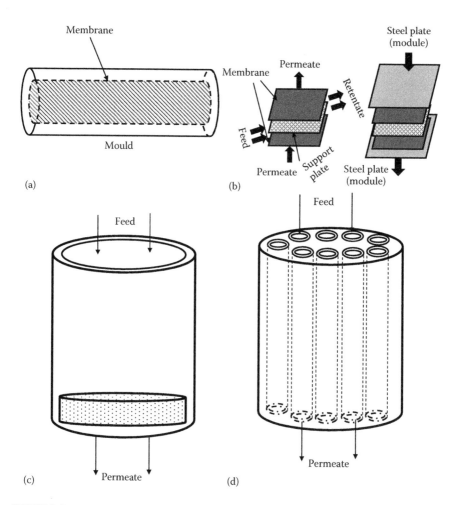

FIGURE 2.2
Schematic representation of different membrane modules. (a) Tubular. (b) Rectangular/flat sheet. (c) Cylindrical. (d) Monolith.

fiber-reinforced fabric pipe. The tube is surrounded within an outer housing or shell having a permeate outlet. Feed is introduced through the tube under pressure. Permeate passes through the membrane and pipe into the interior of the housing and leaves through the permeate outlet. Basically, a tubular module resembles a shell and tube heat exchanger. In some cases, a single tube is used, while in others the module may include a series of tubes arranged axially parallel to one another.

Tubular modules can be operated with simple pretreatment of feed. These modules can be cleaned mechanically by pushing a sponge ball through the tubes. The advantage of this module compared to other modules is minimum membrane fouling due to high feed flow rate.

2.4.2 Rectangular/Flat Sheet

These modules are quite similar to plate and frame filter presses with a series of flat sheet rectangular ceramic membranes. Mostly, rectangular modules are used for accommodating single flat sheet membranes. The membranes are fitted within the module, covered by another part of the module (top section), and sealed by gaskets around the edge of the module to prevent leakage. Feed is generally introduced from the top of the module housing and permeate passes through the bottom the module (Figure 2.2b).

In this type of module, concentration polarization is very low due to the close arrangement of membranes, especially for plate-frame flat sheet modules. Cleaning is also easy for this process.

2.4.3 Cylindrical

Cylindrical modules resemble tubular modules, but the housing contains a tube only, unlike a shell and tube configuration. Basically, these types of modules are used for fitting membranes in cylindrical shapes. In some cases, circular disk shaped membranes can also be suited for cylindrical modules. Cylindrical modules can be operated as dead-end filtration (vertically oriented) for circular disk membranes. For dead-end modules, feed is introduced at the top of the housing (Figure 2.2c) and permeate is collected from the bottom of the housing.

The problem with this module is the maximum probability of pore blockage of the membrane due to high concentration polarization if the membrane is dead end. This problem can be minimized if the membrane is cylindrical in nature due to the cross-flow pattern of the feed.

2.4.4 Monolithic

A monolithic multichannel cross-flow filtration module (Figure 2.2d) includes a porous monolithic body that defines a number of flow channels existing in the body and covering from an upstream inlet or feed side to a downstream outlet or permeate side. Porous channel walls surround each of the group of flow channels. The range of flow channels has a channel hydraulic diameter ≤1.1 mm. Moreover, the porous body includes a networked pore structure of interconnected pores forming tortuous fluid paths or conduits. The tortuous paths formed by the porous body provide a flow path for directing filtrate to be separated from a process stream to an exterior surface of the body.

2.5 Ceramic Membrane Fabrication Techniques

Ceramics can be fabricated by a variety of methods, depending on whether the starting materials are dry or semidry, in the form of slurry or plastic

TABLE 2.1

Traditional Ceramic Fabrication Methods

State of Raw Material Mixture	Fabrication Methods
Plastic mass	Extrusion; injection molding
Slurry (wet basis)	Slip casting; tape casting
Dry or semidry	Die compaction; isostatic pressing

mass, and involve a gaseous, liquid, or solid phase. The objective is the fabrication of a ceramic with the desired shape (tubular, rectangular, circular, or monolithic) and the desired microstructure from suitable raw materials. In the following sections, the main features of processing steps involved in the conventional ceramic forming methods and their advantages and disadvantages are briefly explained. On the basis of form or state of the starting material mixture, fabrication methods can be divided into different classes, summarized in Table 2.1.

Moreover, on the basis of gaseous phase and liquid phase, ceramic fabrication can be categorized as chemical vapor deposition (CVD) (gaseous), directed metal oxidation (gas–liquid), reaction bonding (gas–solid/liquid–solid), and sol-gel (liquid) processes.

2.5.1 Extrusion

Extrusion is mainly used in the manufacturing of traditional ceramics to form moist clay–water mixtures into green body with symmetrical cross sections and to obtain excellent morphological characteristics. The method is also suitable for fabricating a variety of advanced ceramics, such as catalyst supports, electrical insulators, and capacitor tubes. The extrusion process (Figure 2.3) can be divided in three steps. Firstly, the stock material is

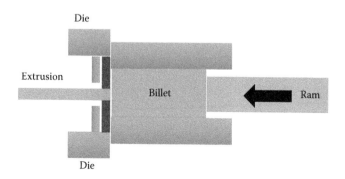

FIGURE 2.3
A schematic of the extrusion process.

heated (for hot or warm extrusion). Secondly, the hot or warm feed is loaded into the container in the press and a dummy block is placed behind it, where the ram then presses on the material to push it out of the die. Finally, the extrusion is stretched in order to straighten it. The extrusion process can be classified into three different segments: hot extrusion, cold extrusion, and warm extrusion.

Hot extrusion is a hot working process as it is done above the material's recrystallization temperature to keep the material from hardening and to make it easier to push the material through the die [41]. Cold extrusion is done at room temperature or near room temperature. The advantages of cold over hot extrusion are the lack of oxidation, higher strength due to cold working, closer tolerances, better surface finish, and fast extrusion speeds if the material is subject to heat loss. Warm extrusion is done above room temperature, but below the recrystallization temperature of the material; the temperatures ranges from 424°C to 975°C. It is usually used to achieve the proper balance of required forces, ductility and final extrusion properties [42].

There are many different variations of extrusion equipment. They vary by four major characteristics:

a. Movement of the extrusion in relation to the ram (If the die is held stationary and the ram moves toward it, then it is called direct extrusion. If the ram is held stationary and the die moves toward the ram, it is called indirect extrusion.)
b. The position of the press, either vertical or horizontal
c. The type of drive, either hydraulic or mechanical
d. The type of load applied, either conventional (variable) or hydrostatic

Advantages include:

- Ability to create very complex cross sections and to provide flexibility to the products
- Efficient melting
- High production volumes
- Excellent surface finish
- Low cost

Disadvantages include:

- Uniform cross-sectional shape only
- Limited complexity of parts

2.5.2 Injection Molding

The manufacturing of ceramic material by injection molding involves four distinct steps (Figure 2.4) [43,44]:

- Feedstock preparation, which includes selection of the powder and the binder, mixing the powder with binder, and production of a homogeneous feed material in the form of granules
- Injection molding
- Debinding (removal of the binder)
- Sintering

The first step in this process is the preparation of feedstock materials. The term *feedstock* refers to a homogeneous pelletized mixture of ceramic powder and a binder (organic or polymeric). The powder and binder are hot mixed above the melting point of the binder elements (for complete burnout or removal) to provide a uniform coating on the powder surface. A homogeneous distribution of powder particles and binder in feedstock is significant as it helps to reduce segregation during the injection molding stage and, later, to obtain isotropic shrinkage after debinding and sintering [45]. Eluding segregation of feedstock constituents is necessary to avoid visual defects, extreme porosity, warping, and flaws in the sintered ceramic body [46]. To obtain an

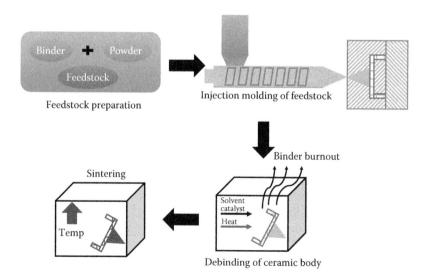

FIGURE 2.4
Flow diagram illustrating injection molding steps for ceramic fabrication.

excellent homogeneous mixture, the binder and the powder are combined in extruders and mixers. The mixture is then pelletized to an appropriate shape for feeding into the molding machine. For this process, thermoplastic polymers that maintain and uphold the shape of the molded part until the last stages of debinding are used as binder. Most commonly used binders are polyvinyl acetate (PVA), polyethylene (PE), polypropylene (PP), polystyrene (PS), polyethylene glycol (PEG), and polymethyl methacrylate (PMMA) [47]. Mechanical methods that involve grinding or milling (comminution) for size reduction of a coarse, granular material are preferable to chemical methods in terms of cost for synthesis of ceramic powders.

The second step is molding the feedstock into the desired shape. Molding of the feedstock is similar to the injection molding of plastics and has the following stages [48,49]:

a. The pelletized feedstock is placed in the hopper of the injection molding machine.

b. The binder in the feedstock is melted by the heating system.

c. The molten material is injected under high pressure ($\geq$60 MPa) into the mold cavity, which is mounted in the clamping unit. The feedstock must have low enough viscosity that it can flow into the die cavity under pressure.

d. The mold remains closed while cooling channels in the die extract heat from the molten feedstock and solidify the polymer to preserve the molded shape.

e. After solidification of the binder, the nozzle of the injection die is pulled away from the mold by moving the injection unit. The clamping unit opens and the molded part is ejected by the ejector system of the machine.

f. The green part is removed from the mold. Due to the fragile nature of most green parts, the removal process is done by hand or by a robotic system in order to prevent shocks or impacts that could deform or even break the molded part.

The third step is debinding, in which the organic binder must be removed without disrupting the molded powder particles. Binders have to be removed completely from the ceramic body before sintering, since carbon residues can affect the sintering process and reduce the quality of the final product. Moreover, precaution should be taken during binder removal to minimize defects in final products. The defects can be produced by inadequate debinding such as bloating, blistering, surface cracking, and large internal voids.

The final step is sintering, a thermal treatment that transforms ceramic powders into bulk materials with improved mechanical strength and excellent morphological properties (porosity, pore size) with developed microstructure.

Advantages include:

- Fast production
- Material and design flexibility
- Low labor cost
- Low waste

Disadvantages include:

- High initial tooling cost
- Restricted part design
- Difficult to price accurately

2.5.3 Pressing or Compaction

Die compaction (uniaxial pressing) and isostatic pressing are frequently used for the compaction of dry powders, which typically contain <2 wt% moisture or water molecules, and semidry powders, which contain approximately 5 wt% to 20 wt% water [50,51]. Die compaction is one of the most widely used processes in the ceramic industry for the production of simple shapes with appropriate dimensions (Figure 2.5). Isostatic pressing provides more homogeneous packing density to the raw materials during formation of ceramics with complex shapes and with larger height-to-diameter ratios.

In the die compaction method, a granular or powder material undergoes simultaneous punching via uniaxial compaction and shaping in a rigid and strong die. The entire process consists of three steps: filling the die, powder compaction, and ejection of the compacted powder. Depending upon the punches and relative motion of the die, the process can be categorized as three modes: (1) single action, (2) double action, and (3) floating die. In the single-action mode, bottom punch and die are fixed and the top punch moves, whereas in double-action mode, both punches move, but the die is fixed. The double-action mode compaction provides better packing homogeneity. In the floating-die mode, the bottom punch is fixed, but both the die and top punch move.

During die compaction, it is important to understand the flow behavior of the raw material to get hold of effective die filling, fast pressing rates, and a reproducible ceramic body. Moreover, it is often necessary to granulate fine powders using a spray drying process as they do not flow very well and are difficult to compact homogeneously. The granule characteristics are dependent on particle size distribution of the initial powder, the degree of flocculation of the slurry, the type of additives, and the spray drying conditions.

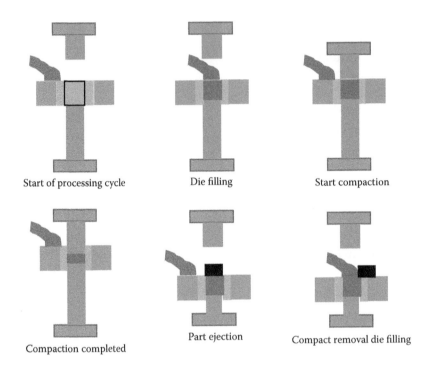

Start of processing cycle   Die filling   Start compaction

Compaction completed   Part ejection   Compact removal die filling

FIGURE 2.5
A schematic illustrating different stages of the die compaction method.

The vital granule characteristics are particle packing and its homogeneity, particle size, shape and size distribution, hardness, and surface friction.

During die filling, flow of granules is controlled by a wide distribution of particle size; uniform particle size is also considered to be an important factor because it will affect the packing homogeneity of the ceramic. Filling uniformity also depends on the ratio of the granule size to the die diameter. Wider dies lead to a higher overall packing density in the compaction because the packing density near the die walls is higher.

Compaction is done after die filling. This operation is divided into two stages. The first stage includes reduction of the large voids by rearrangement of the granules, whereas in the second stage, the small voids are reduced by deformation of the granules. The compaction of particles depends on the quantity and nature of agglomerates. For a particle having no agglomerates but large voids, the compaction involves only one stage (i.e., sliding and rearrangement to reduce the voids). If the powder contains low-density weak agglomerates (presence of both large and small voids), then compaction takes place via two stages. Firstly, rearrangement and sliding take place to reduce the larger voids and, secondly, fracture of agglomerates and further rearrangement and sliding occur to reduce the smaller voids.

Advantages include:

- Simple and inexpensive experimental equipment
- Easy to handle
- Consumes less time

A *disadvantage* is that high die-wall friction generates density variation, leading to an inhomogeneous mixture.

Isostatic compaction provides uniform hydrostatic pressure to the powder contained in a flexible rubber container. There are two types of isostatic pressing or compaction: wet-bag pressing and dry-bag pressing, illustrated in Figure 2.6. Wet-bag pressing is generally used for manufacturing of complex shapes and large sizes of ceramics. Dry-bag pressing is easier to handle than wet-bag pressing.

Steps in wet-bag pressing include:

- A flexible rubber mold is filled with the powder, submerged in a pressure vessel filled with oil, and pressed.
- After pressing, the mold is removed from the pressure vessel.
- A rigid, green ceramic body is then retrieved.

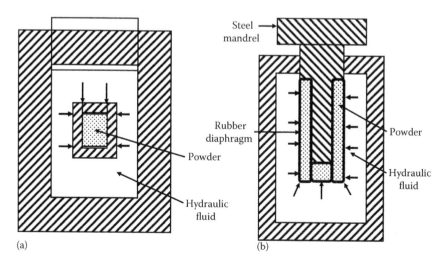

FIGURE 2.6
Two types of isostatic pressing: (a) wet-bag pressing and (b) hot-bag pressing.

Steps in dry-bag pressing include:

- The mold is fixed in the pressure vessel and need not be removed.
- The pressure is applied to the powder kept between a thick rubber mold and a rigid core.
- After relieving pressure, the powder compact is removed from the mold.

Advantages include:

- Powder is compacted with the same pressure in all directions.
- High and uniform density can be achieved.
- Powder densification can be reproduced.
- Production time is short.

Disadvantages include:

- Process is costly.
- This is a near-net shape process technology suited to parts with much wider tolerance requirements.

2.5.4 Slip Casting

Slip casting, one of the most commonly used and time-consuming traditional ceramic forming methods, offers a route for the production of complex shapes with homogeneous particle packing. The term "slip" indicates a mixture of suspended ceramic particle and organic liquid that forms a powder suspension or slurry. The slurry (or slip) contains clay powder, dispersing agent, additives (binders and plasticizers), and water. The entire process can be divided into various stages, such as slurry preparation, mold filling, forming of the cast, slip draining and partial drying, plugging and separating, and final drying. As shown in Figure 2.7, when a well-mixed slurry is poured into a porous mold, a solvent of suspension or slurry is extracted into the pores of the mold through capillary suction pressure, which draws the liquid from the slurry into the mold. Therefore, the slip particles are consolidated on the surface of the mold to form a gel layer (cake). The formation of a consolidated layer (cast) must occur quickly so that the particle suspension is unable to penetrate into the pores of the mold. After a sufficient thickness of the gel layer has formed, the excess slip is poured out and the mold and cast are allowed to dry. Generally, the cast shrinks away from the mold during drying and can easily be removed. After complete drying, the cast is heated to remove any organic additives and then sintered to produce the final product.

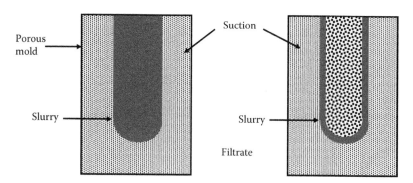

FIGURE 2.7
A general representation of the slip casting method.

Advantages include:

- Low capital investment
- Highly homogeneous slurries
- A wide variety of complex shapes

Disadvantages include:

- Low production rate
- Lower dimensional precision
- Differential shrinkage
- Less mold toughness

2.5.5 Tape Casting

Tape casting, sometimes referred to as the *doctor blade* process, is a casting process used in the manufacture of large thin ceramic tapes (wide-flat layers) from ceramic slurry. Figure 2.8 illustrates the basic principle of the tape casting process, which consists of a stationary casting knife, a reservoir for holding the slurry or powder suspension, a moving carrier, and a drying zone. In this process, the slurry, consisting of the ceramic powder in a solvent, with the addition of binders, plasticizers, and dispersants, is cast over a surface covered with a removable sheet of plastic or paper using a carefully controlled doctor blade. The flexible ceramic tape can be stored on take-up rolls or uncovered from the carrier surface and cut into the desired lengths for further processing. During the subsequent drying process, the dispersing liquid is evaporated, whereupon the tape thickness is reduced. The thickness of the cast layer is controlled by the gap between the casting knife blade (doctor blade) and the carrier. Sometimes, a double doctor blade is found to be useful to maintain a uniform thickness over a long length of tape. For the

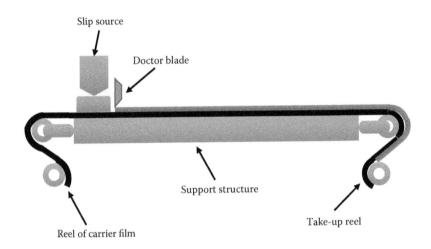

Slip source

Doctor blade

Support structure

Take-up reel

Reel of carrier film

FIGURE 2.8
Schematic diagram of the tape casting process.

production of long ceramic tapes, the blade is kept stationary and the surface moves, whereas for the production of short ceramic flat layers, the blade moves over the stationary surface.

The preparation of slurry is a critical stage in the tape casting process. Some general rules can be considered during the preparation of a tape casting slurry: (1) the ratio between organic components and ceramic powder must be as small as possible; (2) a minimum amount of solvent must be fixed to maintain a homogeneous slurry; (3) the quantity of dispersant must be the least necessary to ensure stability of the slurry; and (4) the plasticizer-to-binder ratio must be adjusted to make the tape flexible, resistant, and easy to release [52].

The casting speed is another factor influencing the process during manufacturing of ceramic tapes. This speed is mainly determined by either a continuous or batch casting process. For a continuous process, the speed is determined by the thickness of the tape, the length of the casting machine, and the volatility of the solvent [40].

The *advantage* is its purity. *Disadvantages* include higher cost and minimum shrinkage.

2.5.6 Chemical Vapor Deposition

CVD is a ceramic forming process by virtue of deposition of all classes of materials (metals, ceramics, etc.) over ceramic via transportation of reactive molecules in the gas phase to a surface at which they chemically react and form a solid film [53–55]. In other words, CVD is a process that has an ability to modify the properties of membrane surfaces by depositing a layer of a different or the same compound through chemical reactions in a gaseous phase surrounding the component at a raised temperature.

A CVD technique includes an arrangement of monitoring and metering (control of gas flow by flow regulators and control valves), a mixture of reactive gas molecules and carrier gases, a reaction chamber, and a scheme for the treatment and disposal of exhaust gases. The gas mixture (reactive gases like metal halides and hydrocarbons in the presence of hydrogen, nitrogen, and argon) is carried into a reaction chamber heated at the desired temperature (Figure 2.9).

Table 2.2 shows some of the important reactions under different temperature ranges used for the manufacturing of ceramics and the application of the final product. The CVD technique can also be classified by several methods based on operating pressure, physical characteristics of vapor, and plasma processing, summarized in Table 2.3.

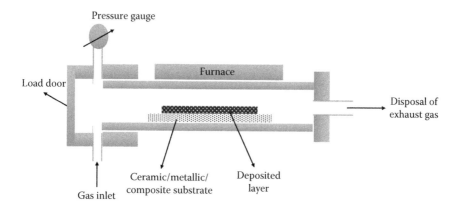

FIGURE 2.9
A unit of chemical vapor deposition (CVD).

TABLE 2.2

Important Chemical Reactions Involved in Ceramic Fabrication Using CVD Process

Reaction	Temperature (°C)	Applications
Oxidation		
$TiCl_4 + O_2 \rightarrow TiO_2 + 2Cl_2$	900–1200	Films/layers for electronic devices
Co-reduction		
$SiCl_4 + 2CO_2 + 2H_2 \rightarrow SiO_2 + 4HCl + 2CO$	800–1000	Films for electronic devices, optical fibers
Hydrolysis		
$SiCl_4 + 2H_2O \rightarrow SiO_2 + 4HCl$	500–1000	Films for electronic devices, optical fibers
Thermal decomposition		
$CH_3Cl_3Si \rightarrow SiC + 3HCl$	1000–1300	Manufacturing of composite

TABLE 2.3

Classifications of Chemical Vapor Deposition Technique

Classification	Method
Operating pressure	Atmospheric pressure CVD (*APCVD*): atmospheric pressure.
	Low-pressure CVD (*LPCVD*): subatmospheric pressures. Reduced pressures tend to reduce unwanted gas-phase reactions and improve film uniformity across the wafer.
	Ultrahigh vacuum CVD (*UHVCVD*): this consists of very low pressure, typically below 10^{-6} Pa.
Physical characteristics of vapor	Aerosol-assisted CVD (*AACVD*): the precursors are transported to the substrate by means of a liquid/gas aerosol, which can be generated ultrasonically. This technique is suitable for use with nonvolatile precursors.
	Direct liquid injection CVD (*DLICVD*): the precursors are in liquid form (liquid or solid dissolved in a convenient solvent). Liquid solutions are injected in a vaporization chamber toward injectors (typically car injectors). The precursor vapors are then transported to the substrate as in classical CVD. This technique is suitable for use on liquid or solid precursors. High growth rates can be reached using this technique.
Plasma methods	Microwave plasma-assisted CVD (*MPCVD*): this is a combination of magnetic field and microwave excitation.
	Plasma-enhanced CVD (*PECVD*): utilizes plasma to enhance chemical reaction rates of the precursors. PECVD processing allows deposition at lower temperatures, which is often critical in the manufacture of semiconductors. The lower temperatures also allow for the deposition of organic coatings, such as plasma polymers, that have been used for nanoparticle surface functionalization.
	Remote plasma-enhanced CVD (*RPECVD*): similar to PECVD except that the wafer substrate is not directly in the plasma discharge region. Removing the wafer from the plasma region allows processing temperatures down to room temperature.

The apparatus used for the CVD process depends upon factors such as the reaction, the reaction temperature and the configuration and temperature of the substrate material, the nature and flow rate of the carrier gas, and the pressure in the reaction vessel (1–15 kPa). The product obtained using CVD can be monocrystalline, polycrystalline, amorphous, or epitaxial.

Advantages include:

- Possibility of high growth rates
- Low fabrication temperature
- Can be deposited with very high purity
- High deposition rate
- Can deposit materials that are hard to evaporate
- Good reproducibility
- Can grow epitaxial films

Disadvantages include:

- Temperatures are high.
- Precursors need to be volatile at near room temperatures.
- CVD precursors can be highly toxic, explosive, or corrosive.
- By-products of CVD reactions can be hazardous.
- Some of these precursors, especially the metal–organic precursors, can be quite costly.
- The process costs more.
- Processes are complex.

2.5.7 Directed Metal Oxidation

Fabrication of ceramic via reactions between a gas and liquid is basically impossible because the reaction product generally forms a solid protective coating, thus separating the reactants and successfully preventing the reaction. In this process, a ceramic matrix (e.g., TiN, ZrN, TiC, and ZrC) is formed due to reaction of a molten metal with an oxidant. For the production of composite or ceramic material, a filler material (e.g., particles or fibers) is shaped into a preform of the shape and size desired of the product. A detailed route to fabricate composite by the process (Figure 2.10) is given as follows:

- The first step in the process involves making a shaped preform of the filler material. Preforms consisting of particles are fabricated with traditional ceramic-forming techniques, while fiber preforms are made by weaving, braiding, or laying up woven cloth.
- Next, the preform is put in contact with the parent metal alloy.

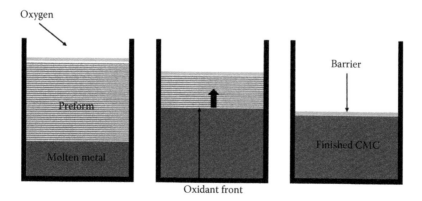

FIGURE 2.10
Fabrication of ceramic-matrix composite (CMC) by direct metal oxidation process.

- A gas-permeable growth barrier is applied to the surfaces of this assembly to limit its shape and size.
- The assembly that is supported in a suitable refractory container is then heated in a furnace.
- The parent metal reacts with the surrounding gas atmosphere to grow the ceramic reaction product through and around the filler to form a ceramic-matrix composite (CMC).
- Capillary action within the growing ceramic matrix continues to supply molten alloy to the growth front.
- The reaction continues until the growing matrix reaches the barrier.
- At this point, growth stops, and the part is cooled to ambient temperature.
- To recover the part, the growth barrier and any residual parent metal are removed.
- Some of the parent metal (5% to 15% by volume) remains within the final composite in micron-sized, interconnected channels.

Advantages include:

- Low shrinkage because near-net shape parts may be fabricated
- Inexpensive and simple equipment
- Inexpensive raw materials
- Good mechanical properties at high temperatures (e.g., creep strength) due to the absence of impurities or sintering aids
- Low residual porosity

Disadvantages include:

- Shrinkage during densification
- Low productivity
- Long fabrication time: 2–3 days

2.5.8 Reaction Bonding

Reaction bonding or reaction forming is mainly used to refer to fabrication routes in which a porous solid preform reacts with a gas or a liquid to yield the desired chemical compound and bonding between the grains. Normally, the process is used to develop near-net shape ceramics due to no or little shrinkage of the preform. Reaction bonding is used for the large-scale production of Al_2O_3 (reaction bonding of aluminum oxide) and Si_3N_4

(RBSN—reaction bonded silicon nitride) [56,57]. The process steps for reaction bonding of alumina (RBAO) are as follows:

- RBAO precursor powders are prepared by attrition milling of Al/Al$_2$O$_3$ mixtures.
- During heat treatment in an oxidizing atmosphere (usually air), the metal phase in RBAO powder compact is fully converted to nanometer-sized oxide crystals, which are sintered.
- They are then bonded with the primary Al$_2$O$_3$ particles.
- The process can be modified in various ways by incorporating metal and ceramic additives to change the final composite composition to accelerate the reaction, to further compensate for the sintering shrinkage.

The process steps for RBSN are as follows:

- Si powder is consolidated by die pressing, isostatic pressing, slip casting, or injection molding to form a shaped article.
- This is then preheated in Ar at 1200°C to develop strength.
- Finally, the component is heated at 1250°C–1400°C (nitrogen at atmospheric pressure) when reaction bonding occurs.
- Nitridation is a mechanism in which new mass that has been added to the body expands into the surrounding pore space.
- As the pore size decreases, the pore channels close off and the reaction effectively stops.

The *advantage* is that a near-net shape ceramic composite can be obtained. *Disadvantages* include:

- The resulting composite may have excessive porosity.
- It is a costly process due to expensive raw material.

2.5.9 Sol-Gel Process

A sol is a suspension of colloidal particles in a liquid or a solution of polymer molecules, whereas a gel is a semirigid mass formed when the colloidal particles form a network either by cross-linking or interlinking of polymer molecules. The sol-gel process is used to fabricate ceramic materials that involve the preparation and gelation of sol [58–60]. Two types of sol-gel processes can be distinguished, depending on whether a sol or solution is used. The two different routes are the particulate gel route (or colloidal; dense particles range from 1 to 1000 nm) (Figure 2.11) and the polymeric gel route (polymer chains; no dense particles greater than 1 nm) (Figure 2.12).

In sol-gel processing, the precursors for the preparation of sol consist of inorganic salt or metal-organic compounds (metal alkoxides). Metal alkoxides, a class of metal-organic compounds having the general formula $M(OR)_X$, where M is a metal of valence X and R is an alkyl group, are the

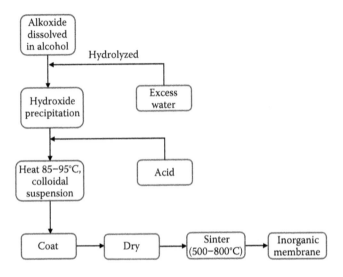

FIGURE 2.11
Flowchart for the sol-gel process via particulate sol route.

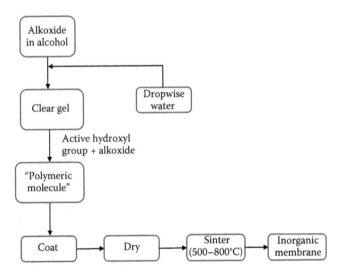

FIGURE 2.12
Flowchart for the sol-gel process via polymer sol route.

most common precursors used in the sol-gel method because they react readily with water. Depending upon the electronegativity of the metal, the methods for the preparation of metal alkoxides are divided into two groups: (1) reaction between metals and alcohols for more electropositive metals and (2) reaction including metal chlorides for relatively less electropositive metals. In addition to this, there are some other processes, such as transesterification, alcohol interchange, and esterification reaction.

In particulate sol, colloidal particles are dispersed in water and peptized with acid or base to produce a sol following three different stages: precipitation of metal alkoxides, peptization (conversion of a precipitate into colloidal sol by shaking it with dispersion medium in the presence of a small amount of electrolyte), and sintering. Gelation can be attained by (1) removal of water from the sol by evaporation to reduce its volume or (2) changing the pH to slightly reduce the stability of the sol. The reactions involved in this process are

$$\text{Precipitation: } Al(OR)_3 + H_2O \rightarrow Al(OH)_3$$

$$\text{Peptization: } Al(OH)_3 \rightarrow \gamma\text{-}Al_2O_3.H_2O \text{ (Bohmite) or } \delta\text{-}Al_2O_3.3H_2O \text{ (Bayenite)}$$

$$\text{Sintering: } \gamma\text{-}Al_2O_3.H_2O \rightarrow \gamma\text{-}Al_2O_3 + H_2O$$

In a polymeric sol route, the reactions involved throughout the process are hydrolysis, polymerization, and cross-linking. Polymerization occurs in three stages:

1. Polymerization of monomers to form particles
2. Growth of particles
3. Linking of particles into chains

Three reactions involved in this process are

$$\text{Hydrolysis: } Ti(OR)_4 + H_2O \rightarrow Ti(OR)_2(OH)_2 + ROH$$

$$\text{Polymerization: } nTi(OR)_2(OH)_2 \rightarrow [-Ti(OR)_2 - O-]_n + H_2O$$

$$\text{Cross-linking: } [-Ti(OR)_2 - O-]_n \rightarrow [-Ti(OH)_2 - O-]$$

Advantages include:

- Highly pure product
- Good chemical homogeneity with multicomponent system
- Lower temperature sintering for ceramic fabrication
- Preparation of ceramics and glasses with novel compositions
- Ease of fabrication for films and fibers

Disadvantages include:

- Expensive starting material
- Conventional drying
- Limited to the fabrication of small articles
- Drying step leading to long fabrication time
- Special handling of raw materials usually required

2.6 Sintering

The heat treatment step in which the dried, loosely bonded (complete removal of binder) green body is converted to a final and useful rigid solid with the required microstructure is termed sintering or firing. In this section, the stages and mechanism of sintering, different mass transports involved during sintering, and different types of sintering methods are outlined. The driving forces for sintering are the curvature of the particle surfaces, an extremely applied pressure, and a chemical reaction. It is necessary during the sintering process to understand the densification steps of a ceramic body and influence of key process parameters such as particle size, temperature, and pressure on it.

2.6.1 Mechanisms and Stages of Sintering

Sintering of polycrystalline materials takes place by diffusion of atoms following definite paths. There are two different mechanisms of sintering in polycrystalline materials: densifying and nondensifying mechanisms. They lead to bonding and growth of necks between the particles to enhance the strength of the material during sintering. Nondensifying mechanisms are categorized as surface diffusion, lattice diffusion from the particle surfaces to the neck, and vapor transport leading to neck growth without densification (mechanisms 1, 2, and 3), as shown in Figure 2.13. Grain boundary

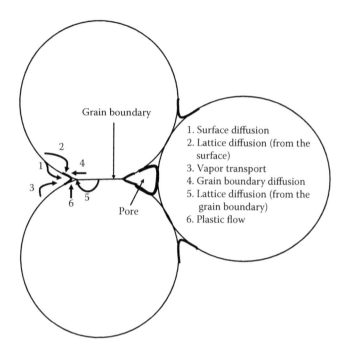

Grain boundary

1. Surface diffusion
2. Lattice diffusion (from the surface)
3. Vapor transport
4. Grain boundary diffusion
5. Lattice diffusion (from the grain boundary)
6. Plastic flow

Pore

FIGURE 2.13
Mass transport mechanism involved in the sintering process.

diffusion and lattice diffusion from the grain boundary to the pore (mechanisms 4 and 5) are referred to as densifying mechanisms. Diffusion from the grain boundary to the pore promotes neck growth as well as densification. Mechanism 6 (i.e., plastic flow by dislocation motion) also leads to neck growth and densification. Nondensifying mechanisms lead to coarsening of the microstructure, thus reducing the driving force for the densifying mechanisms.

Surface diffusion is a general process involving the motion of atoms, molecules, and atomic clusters at solid material surfaces. During this kind of diffusion, surface smoothing, particle joining, and pore rounding occur. Lattice diffusion refers to atomic diffusion within a crystal lattice, which occurs by either interstitial or substitutional mechanisms. Vapor transport leads to neck growth without densification. Grain boundary diffusion occurs along a grain boundary and refers to formation of neck growth. Grain boundary diffusion is faster than volume diffusion in orders of magnitude, meaning that grain boundaries are the pathways by which material is transported through and within rocks of low porosity. Plastic flow states deformation of a material that remains rigid under stresses of less than a certain intensity but that behaves approximately as a Newtonian fluid under severe stresses.

Viscous flow is the dominant mass transport mechanism for amorphous material by which neck growth and densification occur.

The sintering process undergoes three different stages on the basis of change in microstructure and the change in geometry or shape of particles due to sintering. The three stages of sintering are (1) presintering, (2) thermolysis, and (3) densification. Densification can be divided into three stages called the initial stage, the intermediate stage, and the final stage [61]. Some important features of these stages follow.

Features of presintering include:

- Sintering does not begin until the temperature exceeds one-half to two-thirds of the melting temperature of the material.
- Temperature would be sufficient to cause a significant atomic diffusion for solid-state sintering and viscous flow when a liquid flow presents.
- This stage refers to vaporization or removal of water or moisture from the surface of the particle.
- Care should be taken to prevent the precursors from cracks or fractures because of the stresses from the pressure of the vapor evolved or from differential thermal expansion of phases.

Features of thermolysis include:

- This step refers to complete burnout of organic components such as binder, dispersant, etc.
- Binder burnout depends on the composition of the binder material, the flow of the gas surrounding the precursor, microstructure of the organic powder, porosity phases, and dynamic changes in the microstructure.
- Thermochemistry of binder and additives, binder concentration, precursor dimensions and configuration, heating rate, and furnace atmosphere are present.
- Care must be taken because incomplete binder removal and uncontrolled thermolysis may cause defects, deformation, cracks, expanded pores, and distortion.

Features of densification (shrinking of pores) include:

- The initial stage would begin as soon as some degree of atomic mobility is obtained, and during this stage rapid interparticle neck growth occurs by diffusion. The major microstructural change in this stage is neck growth between the particles.
- The intermediate stage begins when the pores have reached their equilibrium shapes dictated by the surface and interfacial tensions. Shrinkage is induced by grain growth and a change in the pore geometry.
- In the final stage the pores will shrink continuously and may finally disappear altogether.

2.6.2 Types of Sintering

Depending upon the composition of raw materials, the wide variety of sintering processes developed to obtain ceramics with desired microstructure are categorized as solid-state sintering, liquid-phase sintering, and reactive sintering.

In solid-state sintering, a porous polycrystalline powder compact is heated at a constant rate to obtain a desired final product. When sintering starts, the solid particles become closer together and form necks, thus increasing the strength of the body due to mass transport to the necks by diffusion. Further increase in sintering temperature leads to a growth in the neck diameter and smoothing of the pore surfaces. Mass transport into the pores leads to further shrinkage and forms a dense ceramic structure. Finally, a point is reached when the larger pores start to break up in individual and isolated pores. During this stage, the grains grow continuously until densification stops. This grain growth leads to the coalescence of neighboring pores, causing an increase in average pore size even though the porosity decreases. At the final stage of densification, when the pores become isolated from the neighboring pores, grain growth becomes more rapid [62].

In many ceramic fabrication processes, the liquid phase (preferably water) is usually used as raw material to promote the sintering process by enhancing the densification rate. Enhanced densification can be obtained from improved rearrangements of the particles and greater mass transport through the liquid phase. Even in ceramics, the liquid is normally formed due to the presence of an additive that forms a eutectic liquid with a small amount of powder on heating. The formed liquid should have a low angle of contact for good wetting of the solid and a low viscosity for fast mass transport through the liquid. Liquid-phase sintering is basically used to fabricate lightweight ceramics like Si_3N_4 and SiC.

Liquid-phase sintering is normally classified in three dominant stages [63]: (1) *rearrangement and liquid redistribution of the liquid*, determined by capillary stress gradient; (2) *solution precipitation*, referring to densification and shape housing of the solid phase; and (3) *final densification*, driven by residual porosity in the liquid phase.

In polyphase ceramics, sintering can be accompanied by chemical reactions, termed reactive sintering—for example, fabrication of cutting tools (advanced ceramics) using aluminum nitride:

$$AIN + TiO_2 \rightarrow Al_2O_3 + TiN \text{ (under } N_2 \text{ atmosphere)}$$

The advantage over direct sintering of the final products is that TiN, a conducting material, forms a continuous network through reactive sintering, allowing machining through electroerosion.

2.6.3 Different Sintering Processes

2.6.3.1 *Microwave Sintering*

Since 1970, there has been growing interest in the use of microwaves for heating and sintering more complex ceramics, rather than using conventional heating by electrical furnaces [64]. The difference between microwave sintering and conventional furnace heating is the mechanism of generation of heat. In microwave sintering, the heat is generated internally by interaction of microwaves with the atoms, ions, and molecules of the material. In a furnace, the heat is generated through the heating coil surrounded at the exterior wall of the furnace and then directed through the material from the wall of the furnace (Figure 2.14).

Microwave heating generates heat within the material first and then heat spreads through the entire volume. The ceramic specimen usually contains loose, nonconducting powder and is placed within a microwave cavity. The materials couple with microwaves and first absorb the electromagnetic energy volumetrically, and then transform into heat. A simple consumer microwave can also be used if the ceramic body is correctly insulated. Uniform heating through the ceramic body can be dependent on the shape of the body [65] and the microwave frequency [66]. High microwave frequency tends to heat the exterior of the sample more than the interior. There are two specific mechanisms of interaction between materials and microwaves: (1) dipole interactions and (2) ionic conduction.

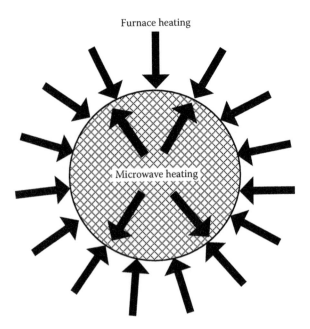

FIGURE 2.14
Heating patterns in conventional furnace heating and microwave furnace heating.

Both mechanisms require effective coupling between components of the target material and the rapidly oscillating electrical field of the microwaves. Dipole interactions occur with polar molecules. The polar ends of a molecule tend to align themselves and oscillate in step with the oscillating electrical field of the microwaves. Collisions and friction between the moving molecules result in heating. Broadly, the more polar a molecule is, the more effectively it will couple with (and be influenced by) the microwave field. Three kinds of materials can be achieved according to the interaction with microwaves: (a) transparent, (b) opaque (conductor), and (c) absorber (Figure 2.15).

The advantages of using microwave heating and sintering ceramic over the conventional sintering include:

- Enhanced diffusion processes
- Reduced energy consumption
- Very rapid heating rates and considerably reduced processing times
- Decreased sintering temperatures
- Improved physical and mechanical properties
- Simplicity
- Unique properties
- Lower environmental hazards

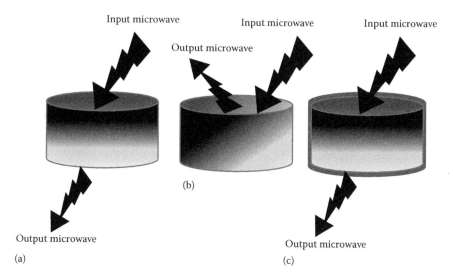

FIGURE 2.15
Generation of heat inside a ceramic body according to the interaction with microwaves. Microwaves can be transmitted, absorbed, or reflected depending on the electrical and magnetic properties of the material. (a) Transparent material through which microwaves can penetrate completely, behaves as low loss insulator. (b) Opaque material that acts as a conductor. Penetration of microwave through this type of material is impossible as microwave totally reflects from the body. (c) This type of material absorbs microwave partially to total.

2.6.3.2 Plasma-Assisted Sintering

Spark plasma sintering (SPS) is a newly developed sintering process to enhance the heating rates during sintering; it makes use of a microscopic electric discharge (DC current pulse) between the particles contained in a graphite die, under pressure (Figure 2.16). The high heating rates are caused by spark discharges generated in the void spaces between the particles. The SPS approach is suitable in producing dense ceramics from nanoscale powders [67,68].

2.6.3.3 Pressure-Assisted Sintering

There are three primary approaches of pressure-assisted sintering: hot pressing, hot isotactic pressing, and sinter forging. The advantage of using pressure-assisted sintering techniques is to enhance the densification rate (lattice diffusion) relative to coarsening rate (surface diffusion), thus providing high density with fine grain size.

Hot pressing: Hot pressing is a high-pressure, low-strain-rate powder metallurgy process for forming a powder or powder compact at a temperature high enough to induce sintering and creep processes. Usually this technique is used to obtain dense ceramics (e.g., synthesis of boron carbide, titanium diboride, and sialon) with simultaneous application of pressure and temperature. A schematic of a standard hot pressing process is shown in Figure 2.17.

There are three distinct heat distribution mechanisms for hot pressing (Figure 2.18): conventional inductive heating, indirect resistance heating, and direct hot pressing.

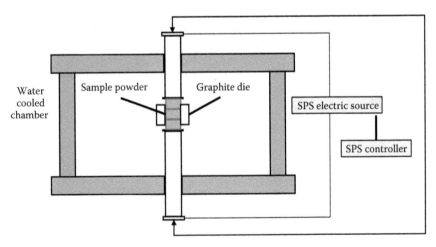

FIGURE 2.16
Schematic of spark plasma sintering equipment.

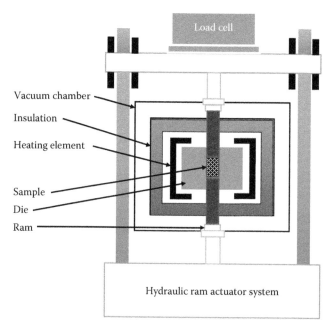

FIGURE 2.17
Schematic of a standard hot-pressing process.

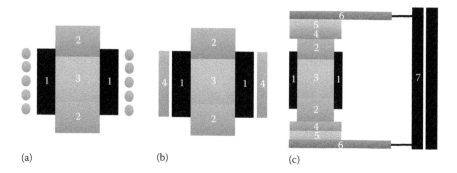

FIGURE 2.18
Heat distribution mechanism for hot pressing method: (a) conventional inductive heating, 1: indicates graphite and/or steel mold, 2: indicates cylinders applying pressure, 3: indicates ceramic powder, and 4: indicates induction coil; (b) indirect resistance heating, 1: indicates graphite and/or steel mold, 2: indicates cylinders applying pressure, 3: indicates ceramic powder, and 4: indicates graphite heating rod; and (c) direct hot pressing, 1: indicates graphite and/or steel mold, 2: indicates cylinders applying pressure, 3: indicates ceramic powder, and 4: indicates upper and lower punches, 5: indicates upper and lower protection plate, 6: indicates pyrometer, and 7: indicates pulsed current.

Facts about inductive heating include:

- Heat is produced within the mold when it is subjected to a high-frequency electromagnetic field generated by using an induction coil coupled to an electronic generator.
- A mold is made of either graphite or steel, and pressure is applied by one or two cylinders onto the punches.
- The mold is positioned within the induction coil.
- The magnetic field can penetrate the mold only 0.5 to 3 mm.
- From there on, the heat has to be "transported" into the mold by the thermal conductivity of the mold material.
- Limitations include improper alignment, difficulty obtaining uniform heat distribution, and costs due to use of high-frequency generator.

Facts about indirect resistance heating include:

- The mold is placed in a heating chamber.
- The chamber is heated by graphite heating elements heated by electric current.
- The heat is then transferred into the mold by convection.
- The electrical energy heats the heating elements that then heat the mold in a secondary manner.

Facts about direct hot pressing include:

- When applying a standard (unpulsed) AC or DC current, it is often referred to as fast direct hot pressing (FAST DHP) or rapid hot pressing (RHP).
- The mold is directly connected to electrical power.
- The resistivity of the mold and the powder part generates the heat directly in the mold, resulting in very high heating rates.
- This leads to significant increase in the sintering activity, lowering the threshold sintering temperature and pressure.
- It is independent of the intrinsic property of the mold material (i.e., its thermal conductivity).

Hot isostatic pressing (HIP): Developed around 1955, hot isostatic pressing (Figure 2.19) is a process that subjects a material simultaneously to high temperature and high gas pressure [69]. This type of technique improves materials' mechanical properties and workability by reducing the porosity

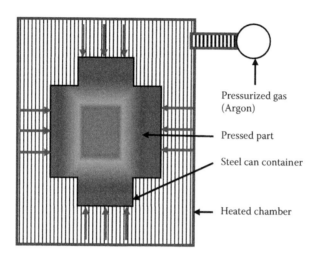

Pressurized gas
(Argon)

Pressed part

Steel can container

Heated chamber

FIGURE 2.19
A general process scheme for hot isostatic pressing.

of a metal and increases the density of many ceramic materials. Pressures up to 300 MPa can be applied with the temperature of an operation as high as 2200°C. The processing time can last up to 4 hours depending upon the material and its size. HIP consists of a pressure vessel, high-temperature furnace, pressurizing system, controls, and auxiliary systems such as material handling, vacuum pumps, and metering pumps. The pressure vessel or container is made of a high-melting-point low alloy steel to obtain high fatigue and creep resistance. Its function is to distribute the heat uniformly throughout the powder while applying gas pressure on all the sides. Generally, a high-temperature inert gas (argon, helium, or nitrogen) is used as a pressurizing medium to ensure uniform pressure distribution throughout the ceramic body. The pressurizing gas is let into the vessel and then a compressor is used to raise the pressure to the desired level. The furnace is then started and both temperature and pressure are increased to a necessary value. Furnaces are radiation or convection types of heating furnaces with graphite or molybdenum heating elements. Nichrome is also used. In this process, the total pressure to the ceramic applied is considered as the addition of pressure due to heating and applied pressure. During HIP, the pores are closed by flow of matter by diffusion and creep, but also bonded across the interface to form a continuous material.

Sinter forging: Sinter forging is quite similar to hot pressing, but without keeping the sample in a die (also known as electrosinter forging [ESF]). This method has been extensively used to produce bismuth titanate with aligned and elongated grain microstructures [70]. Electrosinter forging is obtained by inserting loose, binderless powders into the automatic dosing system, or by manually insertion in the mold. The automatic procedure applies a prepressure onto the powders to ensure electrical contact; hence, it superimposes an intense electromagnetic pulse with a mechanical pulse. The two pulses last 30 to 100 ms. After a brief holding time, the sintered component is extracted by the lower plunger and pushed out by the extractor to leave room for the next sintering. Each sintering round lasts less than one second and is carried out entirely in air.

2.7 Summary

The information related to low-cost ceramic membranes and their fabrication techniques provided in this chapter is summarized as follows:

- Introduction and importance of low-cost ceramic membrane with applications—why to choose a low-cost membrane and how the manufacturing cost of a ceramic membrane can be reduced
- Functions of ceramic membranes—how ceramic membranes behave with different factors or parameters such as raw materials, membrane shapes, and modules; the effect of fabrication techniques and coating techniques on membrane behavior; and the manufacturing cost of ceramic membranes, which is the most promising challenge in ceramic fabrication.
  - Membrane modules
    - Tubular—there is minimum membrane fouling due to high feed flow rate.
    - Rectangular/flat sheet—concentration polarization is much less due to the close arrangement of membranes, especially for plate-frame flat sheet modules. Cleaning is easy.
    - Cylindrical—there is less blockage of the pore if the membrane is cylindrical in shape.
    - Monolith—this is a networked structure.
- Forming of low-cost ceramic membrane—influence of major chemicals, pore-formers, and additives such as binders, plasticizers, solvents, and dispersants as raw materials
  - Major chemicals—natural raw clay, fly ash, cordierite powder, apatite powder, dolomite, calcite, feldspar, and kaolin
  - Pore-formers—sodium carbonate, calcium carbonate, starch, sawdust
  - Additives—solvents (aqueous and organic), dispersants (inorganic acid salts and surfactants), binders (organic and inorganic), plasticizers (dry and wet phases), lubricants (steric acid, stearates), and liquid wetting agents

- Fabrication techniques—conventional fabrication methods are as follows:
  - Extrusion
  - Injection molding
  - Pressing or compaction
  - Slip casting
  - Tape casting
  - Chemical vapor deposition
  - Directed metal oxidation
  - Reaction bonding
  - Sol-gel process
- Sintering—stages, steps (presintering, thermolysis, and densification) and mechanisms; types of sintering
  - Microwave sintering—simple, less energy consumption
  - Plasma-assisted sintering—suitable in producing dense ceramics from nanoscale powders
  - Pressure-assisted sintering—high densification rate
    - Hot pressing
    - Hot isostatic pressing
    - Sinter forging

References

1. N. Saffaj, M. Persin, S.A. Younssi, A. Albizane, M. Bouhria, H. Loukili, H. Dacha, A. Larbot, Removal of salts and dyes by low $ZnAl_2O_4$–TiO_2 ultrafiltration membrane deposited on support made from raw clay. *Separation and Purification Technology*, 47 (2005) 36–42.
2. N. Saffaj, M. Persin, S.A. Younsi, A. Albizane, M. Cretin, A. Larbot, Elaboration and characterization of microfiltration and ultrafiltration membranes deposited on raw support prepared from natural Moroccan clay: Application to filtration of solution containing dyes and salts. *Applied Clay Science*, 31 (2006) 110–119.
3. Y. Dong, X. Liu, Q. Ma, G. Meng, Preparation of cordierite-based porous ceramic micro-filtration membranes using waste fly ash as the main raw materials. *Journal of Membrane Science*, 285 (2006) 173–181.
4. N. Saffaj, S.A. Younssi, A. Albizane, A. Messouadi, M. Bouhria, M. Persin, M. Cretin, A. Larbot, Preparation and characterization of ultrafiltration membranes for toxic removal from wastewater. *Desalination*, 168 (2004) 259–263.
5. S. Masmoudia, A. Larbot, H. E. Feki, R. B. Amara, Elaboration and characterization of apatite based mineral supports for microfiltration and ultrafiltration membranes. *Ceramics International*, 33 (2007) 337–344.

6. J. Zhou, X. Zhang, Y. Wang, A. Larbot, X. Hu, Elaboration and characterization of tubular macroporous ceramic support for membranes from kaolin and dolomite. *Journal of Porous Materials*, 17 (2010) 1–9.
7. J. Liu, Y. Dong, X. Dong, S. Hampshire, L. Zhu, Z. Zhu, L. Li, Feasible recycling of industrial waste coal fly ash for preparation of anorthite-cordierite based porous ceramic membrane supports with addition of dolomite. *Journal of the European Ceramic Society*, 36 (4) (2016) 1059–1071.
8. M.-M. Lorente-Ayzaa, S. Mestre, M. Menendez, E. Sanchez, Comparison of extruded and pressed low cost ceramic supports for microfiltration membranes. *Journal of the European Ceramic Society*, 35 (2015) 3681–3691.
9. F. Bouzerara, A. Harabi, S. Achour, A. Larbot, Porous ceramic supports for membranes prepared from kaolin and doloma mixtures. *Journal of the European Ceramic Society*, 26 (2006) 1663–1671.
10. S. Bose, C. Das, Preparation and characterization of low cost tubular ceramic support membranes using sawdust as a pore-former. *Materials Letters*, 110 (2013) 152–155.
11. S. Bose, C. Das, Role of binder and preparation pressure in tubular ceramic membrane processing: Design and optimization study using response surface methodology (RSM). *Industrial & Engineering Chemistry Research*, 53 (2014) 12319–12329.
12. S. Bose, C. Das, Sawdust: From wood waste to pore-former in the fabrication of ceramic membrane. *Ceramics International*, 41 (3) part A (2015) 4070–4079.
13. A. Prüss-Üstün, R. Bos, F. Gore, J. Bartram, Safer water, better health: Costs, benefits and sustainability of interventions to protect and promote health. World Health Organization, Geneva, 2008.
14. M. Pritchard, T. Mkandawire, J.G. O'Neill, Biological, chemical and physical drinking water quality from shallow wells in Malawi: Case study of Blantyre, Chiradzulu and Mulanje. *Physics and Chemistry of the Earth Journal*, 32 (2007) 1167–1177.
15. A. Bottino, C. Capannelli, A. Del Borghi, M. Colombino, O. Conio, Water treatment for drinking purpose: Ceramic microfiltration application. *Desalination*, 141 (2001) 75–79.
16. N. Subriyer, Treatment of domestic water using ceramic filter from natural clay and fly-ash. *Journal of Engineering Studies and Research*, 19 (3) (2013) 71–75.
17. S. Khemakhem, A. Larbot, R. B. Amar, New ceramic microfiltration membranes from Tunisian natural materials: Application for the cuttlefish effluents treatment. *Ceramics International*, 35 (2009) 55–61.
18. Z. Zhu, Z. Wei, W. Sun, J. Hou, B. He, Y. Dong, Cost-effective utilization of mineral-based raw materials for preparation of porous mullite ceramic membranes via in-situ reaction method. *Applied Clay Science*, 120 (2016) 135–141.
19. K. Hua, A. Shui, L. Xu, K. Zhao, Q. Zhou, X. Xi, Fabrication and characterization of anorthite–mullite–corundum porous ceramics from construction waste. *Ceramics International*, 42 (5) (2016) 6080–6087.
20. Q. Lü, X. Dong, Z. Zhu, Y. Dong, Environment-oriented low-cost porous mullite ceramic membrane supports fabricated from coal gangue and bauxite. *Journal of Hazardous Materials*, 273 (2014) 136–145.
21. R.V. Kumar, A.K. Ghoshal, G. Pugazhenthi, Fabrication of zirconia composite membrane by in-situ hydrothermal technique and its application in separation of methyl orange. *Ecotoxicology and Environmental Safety*, 121 (2015) 73–79.

22. S.H. Paiman, M.A. Rahman, Md. H.D. Othman, A.F. Ismail, J. Jaafar, A. Abd Aziz, Morphological study of yttria-stabilized zirconia hollow fibre membrane prepared using phase inversion/sintering technique. *Ceramics International*, 41 (10) Part A (2015) 12543–12553.

23. C.H. Konrad, R. Völkl, U. Glatzel, A novel method for the preparation of porous zirconia ceramics with multimodal pore size distribution. *Journal of the European Ceramic Society*, 34 (5) (2014) 1311–1319.

24. Z. Zhu, J. Xiao, W. He, T. Wang, Z. Wei, Y. Dong, A phase-inversion casting process for preparation of tubular porous alumina ceramic membranes. *Journal of the European Ceramic Society*, 35 (11) (2015) 3187–3194.

25. J.-H. Ha, S. Zaighum A. Bukhari, J. Lee, I.-H. Song, The preparation and characterizations of an alumina support layer as a free-standing membrane for microfiltration. *Ceramics International*, 41 (10) part A (2015) 13372–13380.

26. W. Qin, B. Lei, C. Peng, J. Wu, Corrosion resistance of ultra-high purity porous alumina ceramic support. *Materials Letters*, 144 (2015) 74–77.

27. Y.H. Wang, X.Q. Liu, G.Y. Meng, Preparation of asymmetric pure titania ceramic membranes with dual functions. *Materials Science and Engineering*, A 445–446 (2007) 611–619.

28. A. Subramanian, H. Kaligotla, An analysis of mass transport fluxes in titania-based mesoporous ceramic matrices. *Powder Technology*, 247 (2013) 270–278.

29. D.E. Koutsonikolas, G. Pantoleontos, S.P. Kaldis, V.T. Zaspalis, G.P. Sakellaropoulos, Preparation and characterization of novel titania-modified ceramic membranes. *Procedia Engineering*, 44 (2012) 908–909.

30. V. Suwanmethanond, E. Goo, P.K.T. Liu, G. Johnston, M. Sahimi, T.T. Tsotsis, Porous silicon carbide sintered substrates for high-temperature membranes. *Industrial Engineering and Chemistry Research*, 39 (2000) 3264–3271.

31. M. Fukushima, Y. Zhou, H. Miyazaki, Y.-I. Yoshizawa, K. Hirao, Y. Iwamoto, S. Yamazaki, T. Nagano, Microstructural characterization of porous silicon carbide membrane support with and without alumina additive. *Journal of the American Ceramic Society*, 89 (2006) 1523–1529.

32. W. Deng, X. Yu, M. Sahimi, T.T. Tsotsis, Highly permeable porous silicon carbide support tubes for the preparation of nanoporous inorganic membranes. *Journal of Membrane Science*, 451 (2014) 192–204.

33. L. Zhu, M. Chen, Y. Dong, C.Y. Tang, A. Huang, L. Li, A low-cost mullite–titania composite ceramic hollow fiber microfiltration membrane for highly efficient separation of oil-in-water emulsion. *Water Research*, 90 (2016) 277–285.

34. D.K. Ramachandran, M. Søgaard, F. Clemens, B.R. Sudireddy, A. Kaiser, Low cost porous MgO substrates for oxygen transport membranes. *Materials Letters*, 69 (2016) 254–256.

35. W.E. Worrall, *Ceramic raw materials*, 2nd ed., Pergamon Press, Oxford, England, 1969.

36. S. Emani, R. Uppaluri, M.K. Purkait, Cross flow microfiltration of oil–water emulsions using kaolin based low cost ceramic membranes. *Desalination*, 341 (2014) 61–71.

37. B.K. Nandi, B. Das, R. Uppaluri, Clarification of orange juice using ceramic membrane and evaluation of fouling mechanism. *Journal of Food Process Engineering*, 35 (3) (2012) 403–423.

38. B.K. Nandi, R. Uppaluri, M.K. Purkait, Treatment of oily waste water using low-cost ceramic membrane: Flux decline mechanism and economic feasibility. *Separation Science and Technology*, 44 (12) (2009) 2840–2869.

39. M-M. Lorente-Ayza, E. Sánchez, V. Sanz, S. Mestre, Influence of starch content on the properties of low-cost microfiltration ceramic membranes. *Ceramics International*, 41 (10) (part A) (2015) 13064–13073.

40. M.H. Rahman, *Ceramic processing*, Taylor & Francis Group, Boca Raton, FL, 2007, ISBN-10: 0-8493-7285-2.

41. E. Oberg, F.D. Jones, H.L. Horton, H.H. Ryffel, *Machinery handbook*, 26th ed., Industrial Press, New York, 2000, ISBN 0-8311-2635-3.

42. B. Vitzur, Metal forming, *Encyclopedia of physical science & technology*, 8th ed., Academic Press, San Diego, 1987, pp. 80–109.

43. A.C. Gonçalves, Metallic powder injection molding using low pressure. *Journal of Materials Processing Technology*, 118 (1–3) (2001) 93–198.

44. B.Y. Tay, N.H. Loh, S.B. Tor, F.L. Ng, G. Fu, X.H. Lu, Characterization of micro gears produced by micro powder injection molding. *Powder Technology*, 188 (3) (2009) 179–182.

45. C. Quinard, T. Barriere, J.C. Gelin, Development and property identification of 316L stainless steel feedstock for PIM and μPIM. *Powder Technology*, 190 (1–2) (2009) 123–128.

46. M. Thornagel, PIM 2010-simulating flow can help avoid mold mistakes. *Metal Powder Report*, 65 (3) (2010) 26–29.

47. P. Thomas-Vielma, A. Cervera, B. Levenfeld, A. Várez, Production of alumina parts by powder injection molding with a binder system based on high density polyethylene. *Journal of the European Ceramic Society*, 28 (4) (2008) 763–771.

48. Arburg powder injection molding (PIM)—Production of complex molded parts from metal and ceramic. GmbH + Co KG, (2009). Lossburg, Germany, Available from http://www.arburg.de/com/common/download/WEB_522785_en_GB.pdf.

49. R.M. German, A. Bose, *Injection molding of metals and ceramics*. Metals Powder Industries Federation, Princeton, NJ (1997), ISBN 187895461X.

50. D. Bortzmeyer, Die pressing and isotactic pressing, in *Materials science and technology*, vol. 17A: Processing of ceramics, part I, R.J. Brook (ed.), VCH, New York, 1996, p. 127.

51. S.J. Glass, K.G. Ewsuk, Ceramic powder compaction. *MRS Bulletin*, 22 (12) (1997) 24–28.

52. D. Hotza, P. Greil, Review: Aqueous tape casting of ceramic powders. *Materials Science and Engineering A*, 202 (1995) 206–217.

53. H.O. Pierson, *Handbook of chemical vapor deposition (CVD): Principles, technology, and applications*, Noyes, Park Ridge, NJ, 1992.

54. M.L. Hitchman, K.F. Jensen, eds., *Chemical vapor deposition: Principles and applications*, Academic Press, London, 1993.

55. A. Sherman, *Chemical vapor deposition for microelectronics*, Noyes, Park Ridge, NJ, 1987.

56. M.D. Snel, G. de With, F. Snijkers, J. Luyten, A. Kodentsov, Aqueous tape casting of reaction bonded aluminium oxide (RBAO). *Journal of the European Ceramic Society*, 27 (1) (2007) 27–33.

57. Y.-M. Chiang, J. S. Haggerty, R. P. Messner, C. Demetry, Reaction-based processing methods for ceramic-matrix composites. *American Ceramic Society Bulletin*, 68 (2) (1989) 420–428.

58. C.J. Brinker, G.W. Scherer, *Sol-gel science*, Academic Press, New York, 1990.
59. L.C. Klein, ed., *Sol-gel technology for thin films, fibers, preforms, electronics and specialty shapes*, Noyes Publications, Park Ridge, NJ, 1988.
60. B.J.J. Zelinski, D.R. Uhlmann, Gel technology in ceramics. *Journal of Physics & Chemistry of Solids*, 45 (10) (1984) 1069–1090.
61. R.L. Coble, Sintering crystalline solids. I. Intermediate and final state diffusion models. *Journal of Applied Physics*, 32 (5) (1961) 787–792.
62. M.F. Yan, Microstructural control in the processing of electronics ceramics. *Material Science and Engineering*, 48 (1981) 53–72.
63. R.M. German, *Liquid phase sintering*, Plenum Press, New York, 1985.
64. D.L. Johnson, Ultra-rapid sintering. *Material Science Research*, 10 (1984) 243–247.
65. A. Birnboim, Y. Carmel, Simulation of microwave sintering of ceramic bodies with complex geometry. *Journal of American Ceramic Society*, 82 (1999) 3024–3030.
66. A. Birnboim, D. Gershon, J. Calame, A. Birman, Y. Carmel, J. Rodgers, B. Levush, Y.V. Bykov, A.G. Eremeev, V.V. Holoptsev, V.E. Semonov, D. Dadon, P.L. Martin, M. Rosen, R. Hutcheon, Comparative study of microwave sintering of zinc oxide at 2.45, 30, and 83 GHz. *Journal of American Ceramic Society*, 81 (1998) 1493–1501.
67. L. Gao, Z. Shen, H. Miyamoto, M. Nygren, Superfast densification of oxide/oxide ceramic composite. *Journal of American Ceramic Society*, 82 (1999) 1061–1063.
68. T. Takeuchi, M. Tabuchi, H. Kageyama, Preparation of dense $BaTiO_3$ ceramics with submicrometer grains by spark plasma sintering. *Journal of American Ceramic Society*, 82 (1999) 939–943.
69. H. Saller, S. Paprocki, R. Dayton, E. Hodge, A method of bonding. Canadian patent #680160 1964.
70. J.S. Patwardhan, M.N. Rahman, Compositional effects on densification and microstructural evolution of bismuth titanate. *Journal of Materials Science*, 39 (2004) 133–139.

3

Characterization Techniques

3.1 Introduction

This chapter elaborates on the understanding of separation performance of ceramic membranes, which is directly associated with membrane morphology and its thermal, mechanical, and chemical stability. Pore size and its distribution, porosity, shape, surface area and texture, particle packing, and density are the key parameters to determine the separation properties for porous ceramic membranes. For dense ceramic membranes, the important parameters to determine are microstructure as well as crystallinity, gas, and liquid permeation study. In addition to this, the separation performances and the stability of the membranes are dependent on the strength of the ceramic membrane, chemical stability (acidic or basic medium), and thermal stability. In general, characterization of ceramic catalytic membranes is more demanding than that of most of the conventional symmetric or asymmetric porous or dense membranes, as the separation layer is a layer of catalyst that may often have a dense and rough surface. Hence, specialized characterization techniques, such as a catalytic activity test, are essential to determine conversion, yield, and selectivity (discussed in Chapter 7).

Therefore, accurate characterization techniques are vital for the development of commercial ceramic membranes. The characterization techniques for ceramic membranes can be classified into three different categories: (1) morphological characterization, (2) permeation study, and (3) determination of thermal, mechanical, and chemical stability. Therefore, in this chapter, characterization techniques such as scanning electron microscopy (SEM), field emission scanning electron microscopy (FESEM), atomic force microscopy (AFM), transmission electron microscopy (TEM), mercury porosimetry, perporometry, thermoporometry, gas adsorption/desorption isotherms, Archimedes' principle, bubble point, and measurement of solute rejection will be discussed for morphological characterization of ceramic membranes. For the parameters related to permeation, characterization techniques such as liquid and gas permeation methods will be introduced. In addition, thermogravimetric analysis (TGA) for determining thermal stability, flexural strength, and fracture toughness, as well as a hardness test for mechanical stability and an acid–alkali test for measurement of chemical stability, are discussed. Some other techniques, such as contact angle measurement for wettability test, zeta potential for surface charge measurement, and x-ray diffraction for the measurement of crystallinity of ceramic membranes will also be discussed in brief.

3.2 Measurement of Thermal Stability

3.2.1 Thermogravimetric Analysis

A material can lose its mass during heating by (1) chemical reactions, (2) release of adsorbed species, or (3) decomposition. All of these show that the material is no longer thermally stable. Generally, TGA is used to estimate the thermal stability of a material. In a desired temperature range, if a species is thermally stable, there will be no observed change in weight or mass. Little or no slope in the TGA trace indicates negligible weight loss of a sample with respect to change in temperature. TGA also provides the temperature regime where the material starts to decompose. Basically, TGA is a process of thermal analysis in which variations in physical and chemical properties of materials are measured as a function of increasing temperature with constant heating rate, or as a function of time with constant temperature and/or constant mass loss. Generally, ceramics melt before their degradation as these are thermally stable over a large temperature range. Thus, TGA is mainly used to study the thermal stability of ceramics and to investigate the characteristics of the raw materials at sintering temperature during fabrication of ceramic membranes.

Operating principle: A TGA analysis is executed by slowly raising the temperature of a sample in a furnace as its weight is measured on an analytical balance. The schematic principle of the TGA measurement is shown in Figure 3.1. The sample is heated under nitrogen atmosphere or oxygen with constant heat rate while the difference of the mass during this process is measured. A mass loss indicates that a degradation of the measured substance has taken place. The weight of the sample is plotted against time or temperature to clarify thermal changes in the material, such as loss of moisture, loss of additives, and, finally, decomposition of the material.

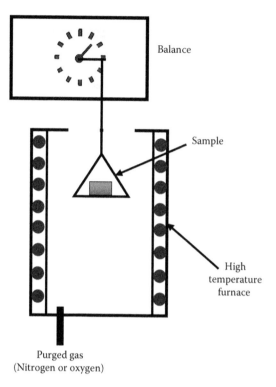

FIGURE 3.1
Schematic presentation of TGA operation.

3.3 Determination of Morphological Characteristics

3.3.1 Physical Methods

3.3.1.1 Mercury Porosimetry

Mercury porosimetry is the most common and widely used method for measuring pore size and pore size distribution in ceramic or inorganic membranes. It should be capable of analyzing powder samples, polymeric thin films, ceramic disks, ceramic tubes, any beads, cylindrical solid samples, etc. It can operate from vacuum to 60,000 psi. Pore diameter from 0.003 to 900 µm can easily be detected using a mercury porosimeter. For high pressure, fluid other than mercury (e.g., water) can also be used. But water will work only for pores that are hydrophobic (i.e., polymer films, polytetrafluoroethylene, polyvinyl chloride). The disadvantages of this method are that high force is required to pass mercury through the small pores, which can damage the membrane structure, and the method considers all the pores, including open

and closed, during the calculation of pore size distribution, which may provide over estimated result.

Operating principle: Mercury is a nonwetting fluid at room temperature for
most ceramic/inorganic materials. A force (gauge) P, in pascals, is applied to
force mercury into a circular cross-section capillary of diameter d, in meters,
given by the equation called the Laplace equation:

$$P = \frac{4\gamma \cos \theta}{d} \text{ or } r_p = -\frac{4\gamma \cos \theta}{P} \tag{3.1}$$

where

γ is the surface tension of the mercury in $N.m^{-1}$
r_p is the pore radius in m
θ is the angle of contact of the mercury on the material being imposed

Figure 3.2 shows the schematic of the process. It is assumed that the pores
of the ceramic membrane are cylindrical, which is not the usual case for most
ceramic membranes, and a morphological constant must be introduced to
obtain accurate results [1].

During the experiment, the largest pores will be occupied by mercury at
a certain minimum pressure. By steadily increasing the force (or pressure)

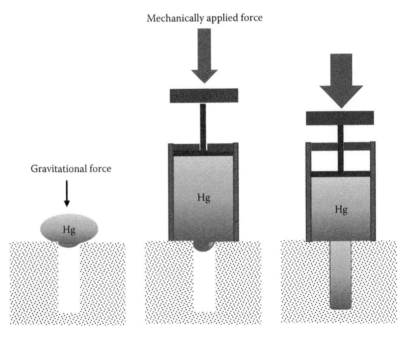

FIGURE 3.2
Schematic presentation of mercury porosimetry.

applied to a porous ceramic membrane immersed in a mercury bath, and by observing the cumulative volume of the mercury imposed for each new applied force, the pore size and its distribution can be evaluated in terms of incremental volume of the mercury as a function of the applied force for a given diameter *d*.

3.3.1.2 Perporometry

Perporometry is a beneficial and convenient characterization technique for the determination of active pores and size distribution for porous ceramic membranes in the range of 2 to 50 nm [2,3]. The only disadvantage of this method is uncertainty of the pore size distribution due to cylindrical pore model deficiency.

Operating principle: The pore size and pore size distribution are determined on the basis of controlled blocking of pores by capillary condensation with condensable vapor, such as cyclohexane or ethanol, combined with a counterdiffused noncondensable vapor like oxygen or nitrogen via the pores, which are not blocked by condensed vapor. Usually, a porous solid is kept in contact with a condensable vapor at low pressure; then the relative vapor pressure (p/p_0) is increased gradually and a monomolecular layer is first formed over the inner surface of the pore wall. With increasing (p/p_0), a multimolecular adsorbed layer forms, followed by capillary condensation of the vapor inside some of the pores causing pore blockage. Capillary condensation can be described by the Kelvin equation:

$$\ln \frac{p}{p_0} = \frac{\zeta \gamma_s V_{mol} \cos \theta}{r_k RT} \tag{3.2}$$

where

ζ is a process parameter
γ_s is liquid–solid interfacial tension in $J.m^{-2}$
V_{mol} is the molar volume of the condensable vapor in $m^3.mol^{-1}$
θ is the contact angle
R is the universal gas constant in $m^3.MPa.K^{-1}.mol^{-1}$
T is the temperature in K
r_k is the Kelvin radius in m

If p/p_0 is high, only the larger pores get filled, but when $p/p_0 = 1$, all the pores are filled.

Experimental arrangement: A membrane is fixed at the middle of a two-compartment cell that is filled with two flowing gas streams: One is an inert gas (i.e., N_2) and the other is a mixture of inert gas, noncondensable gas (O_2), and condensable vapor, such as ethanol (EtOH), at the same pressure. The mixture gas flows at the feed side and the inert gas (i.e., N_2) is introduced at

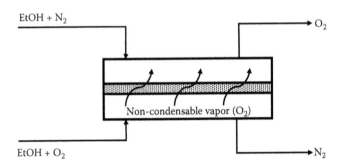

FIGURE 3.3
Process schematic of perporometry.

the permeate side of the membrane. Figure 3.3 illustrates the schematic of the perporometry method. The partial pressure of the condensable vapor is increased first and then reduced to open or close pores of different size ranges. The amount of noncondensable gas (O_2) that diffuses through membrane pores from the permeate side stream will increase the number and size of the pores as pores are opened.

3.3.1.3 Thermoporometry

This method is based on the solid–liquid phase change of a capillary condensate inside a pore of a porous ceramic membrane. Again, the problem in this method is the consideration of dead-end pores during calculation, which is not desirable. By assuming cylindrical pores, a relation between the pore size, r_p in nanometers, and the extent of undercooling, ΔT in kelvins or degrees Celsius (Equations. 3.3 and 3.4) has been derived [4]. The solidification temperature depends on the liquid interface curvature and hence the size of the pore. When the phase change (solidification or melting) takes place, heat is required or released that can be measured by differential scanning calorimetry (DSC). The pore size distribution can be calculated using the data obtained from DSC by plotting thermal flux versus change in temperature.

During solidification, the equation is

$$r_p = \left(-\frac{64.67}{\Delta T} \right) + 0.57 \tag{3.3}$$

During melting, the equation is

$$r_p = \left(-\frac{32.33}{\Delta T} \right) + 0.68 \tag{3.4}$$

where

$$\Delta T = T - T_0$$

T is the temperature where the phase transformation is detected when the liquid is detained in the pores

T_0 is the phase transition temperature of the liquid

3.3.1.4 Gas Adsorption/Desorption Isotherms

Gas adsorption is a simple and widely used method in characterization of porous ceramic membranes for determining pore size and pore size distribution (slit, cylindrical, or spherical) by using specific surface area at monolayer coverage. An inert gas (N_2 at 77K) is used for this process at low relative pressure. Liquid nitrogen is also used, as the pore walls of the ceramic membrane can easily be soaked because of the lower (concave) meniscus of the liquid in the pores. The adsorption/desorption isotherm of the inert gas is determined as a function of the relative pressure (p/p_0). The smallest pores will be filled first by liquid nitrogen at low pressure. With further increase in pressure, larger pores will be filled. When the pressure reaches the saturation point, all pores will be filled. The total pore volume can also be determined by the quantity of gas absorbed close to the saturation pressure. The pore size can be evaluated by the Kelvin equation (Equation 3.2). The distributions of pore volume as a function of pore radius and the distributions of specific surface area can be obtained by using the BJH (Barrett–Joyner–Halenda) method and BET (Brunauer–Emmett–Teller) techniques, respectively. The disadvantage of this method is the consideration of dead-end or closed pores during calculation, which may give ambiguous results.

3.3.1.5 Archimedes' Principle

Volumetric porosities of ceramic membranes can be determined based on gravimetric analysis of water entrapped in the pores of the membrane walls.

Operating principle: The sintered ceramic membranes are first measured in dry condition and then immersed in deionized water for 24 h. The water-soaked membranes are then dried with the help of tissue paper and weighed. The volumetric porosity is calculated from the following equation [5,6]:

$$\varepsilon = \frac{\text{void volume}}{\text{bulk volume}} = \frac{(W_2 - W_1)/\rho_{H_2O}}{V_m} = \frac{(W_2 - W_1)}{V_m \rho_{H_2O}} \tag{3.5}$$

where

W_2 and W_1 are the weights of the wet and dry membrane samples

ρ_{H_2O} is the density of the deionized water

V_m is the volume of the ceramic membrane

This is the simplest and easiest method to determine membrane porosity. But the problem is that the obtained porosity values are found to be slightly higher compared to the values calculated from other standard permeation tests, as water remains intact in the pore interiors and pore walls because of the swelling or hydrophilic nature of ceramic membranes. It is recommended to compare the results with other conventional techniques using statistical study [7].

3.3.1.6 Bubble Point

The bubble point method is the simplest and most used technique for characterizing the size of the largest pore in a ceramic membrane by a stepwise rise in applied pressure. This method is suitable for characterizing macroporous (>50 nm) ceramic membranes.

Operating principle: The basic concept of this method is to measure the minimum pressure required for forcing a gas through a membrane with liquid-filled pores and can be described by the Laplace equation (Equation 3.1). The principle of the bubble point method is shown in Figure 3.4. In this method,

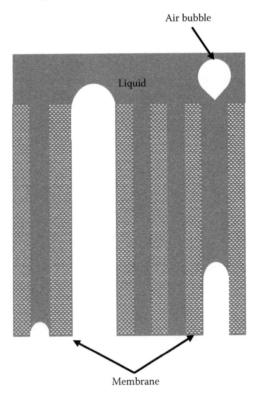

FIGURE 3.4
View of the principle of bubble point.

the liquid-filled membrane acts as a barrier through which flow of air can only occur when the applied pressure reaches the capillary pressure of the largest pores. After raising the pressure over the capillary pressure, the liquid is ejected from the largest pores to permeate the air through it. After reaching the pressure at maximum, the smaller pores expel liquid and allow air to permeate.

Experimental arrangement: A membrane is filled with a wetting liquid and a pressure is applied using gas and increased gradually. The pore diameter is calculated on the basis of the pressure differential required for the flow of air (bubble) to displace the liquid already entrapped within the membrane pores. It is assumed that all the pores are cylindrical in shape. Bubbles are determined either visually or by electronic flow sensors [8]. The surface tension and the contact angle must also be measured during calculation. Pore size distribution can be obtained by measuring the total volume of the displacing fluid as a function of differential pressure.

The problems associated with this method are measurement ambiguity, which will repeatedly underpredict the size of the largest pore; assumption of complete wetting of the pore by the liquid; and the cylindrical shape of the pores is aligned perpendicular to the surface.

3.3.1.7 Measurement of Solute Rejection

Molecular weight cutoff (MWCO) is the empirical method most often used for the determination of complete solute rejection characteristics of porous membranes (reverse osmosis [RO], ultrafiltration [UF], microfiltration [MF], and nanofiltration [NF]). It is defined as the molecular weight of standard monodisperse solute above which molecules are at least 90% rejected by the membrane. Fractional rejection or retention of a solute is defined by Equation 3.6. Basically, this is a size exclusion (size of the pores) based method to measure the rejection of solute. This method also depends upon different types of mixtures of solute and solvent solutions, such as polyethylene glycols (PEGs); dextrans for UF and MF membranes; a mixture of mono- and divalent salts such as sucrose, urea, and alcohols for RO; dialysis membranes; and methanol or dimethyl methylsuccinate (DMMS) for NF membranes [9,10].

$$R_1 = 1 - \frac{C_p}{C_f} \tag{3.6}$$

where C_p and C_f are the solute concentration in permeate and in the feed, respectively (mol.m^{-3}).

The difficulty related to this method is the assumption of the solute rejection, which is based only on molecular weight and may cause an error in analysis. An overestimate of percentage of rejection also may be possible due to surface interaction between the solutes and the membrane surface.

3.3.2 Permeation Methods

3.3.2.1 Liquid Permeation

Usually, morphology of ceramic membranes looks like an assembly of particles with void spaces, which is similar to a layer of packed particles. Evaluation of solvent flux in the membrane module as a function of pressure drop is necessary and can be described by Darcy's law:

$$J_w = L_p \Delta P \tag{3.7}$$

where
J_w is the solvent flux in $m^3.m^{-2}.s^{-1}$
ΔP is the pressure driving force
L_p is the permeability of water, which depends on a number of factors, such as membrane morphology (pore size distribution, porosity, and tortuosity), liquid density, and viscosity

The average pore radius of the membrane pores can be described by the Hagen–Poiseuille equation when the flow of the liquid through the pores is laminar.

$$J_w = \frac{\varepsilon d^2 \Delta P}{32 \mu \tau l_m} \tag{3.8}$$

where
ε is the membrane porosity
d is the pore diameter in m
μ is the liquid viscosity in $kg.m^{-1}.s^{-1}$
τ is the tortuosity of the pores
l_m is the membrane thickness in m

If the flow of the liquid through the pores is turbulent, then the average pore radius of the membrane pores can be described by the Kozeny–Karman equation:

$$J_w = \frac{\varepsilon^3 d_s^2 \Delta P}{180 \mu (1-\varepsilon)^2 l_m} \tag{3.9}$$

where d_s is the equivalent particle diameter. Once membrane porosity, thickness, and tortuosity are measured independently, average pore radius can easily be determined from the slope value by plotting a graph between the solvent flux and pressure drop from experimentally obtained data.

3.3.2.2 Gas Permeation

The gas permeation data are utilized to estimate two vital membrane characteristics—namely, average pore radius (r_g) and effective porosity $\left(\dfrac{\varepsilon}{\tau^2}\right)$ according to the following expression [11]:

$$k = 2.133\left(\frac{r_g}{l}\right)v\frac{\varepsilon}{\tau^2} + 1.6\left(\frac{r_g^2}{l}\right)\frac{1}{\eta}\left(\frac{\varepsilon}{\tau^2}\right)P \tag{3.10}$$

where
  P is the average pressure on the membrane in MPa
  v is the molecular mean velocity of the operating gas (m.min^{-1})
  l is the pore length in m
  τ is the tortuosity factor of the membrane
  η is the viscosity of the gas in Pa.s
  k is the effective permeability factor in m.min^{-1}

Equation 3.10 simply shows a straight line when a graph is plotted between k and P. In this equation, the first term (intercept) and the second term (slope) resemble Knudsen and viscous permeance, respectively. The values of the slope and intercept obtained from the graph are used to evaluate the pore diameter and porosity of the membrane.

The effective permeability factor k is calculated as

$$k = \frac{QP_1}{S\Delta P} \tag{3.11}$$

where
  S is the permeable area of the membrane in m^2
  Q is the volumetric flow rate in m^{3}min^{-1}
  P_1 is the membrane pressure at permeate side in MPa
  ΔP is the transmembrane pressure drop in MPa

The average pore radius (r_g) of the membrane is evaluated using the slope $1.6\left(\dfrac{r_g^2}{l}\right)\dfrac{1}{\eta}\left(\dfrac{\varepsilon}{\tau^2}\right)$ and intercept $2.133\left(\dfrac{r_g}{l}\right)v\dfrac{\varepsilon}{\tau^2}$ of the created graph, expressed as

$$r_g = 1.333\frac{B_1}{A_1}v\eta \tag{3.12}$$

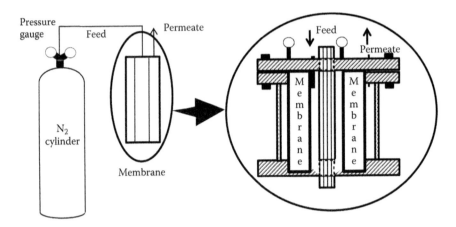

FIGURE 3.5
Experimental setup for gas permeability test.

where B_1 is the intercept and A_1 is the slope. The gas permeance (p) is calculated as

$$p = \frac{Q}{S\Delta P} \qquad (3.13)$$

Experimental arrangement: This is a very simple method and requires a very simple and cheap experimental setup. As shown in Figure 3.5, the setup used for this experiment consists of a tubular cell mounted on a flat rectangular base plate made of perspex [5–7].

The bottom end of the tubular cell is fixed with a flat, rectangular plate and the top end is coupled with two rectangular flat plates by four flanges (black sticks). The membrane is fitted into the hollow sector inside the tubular cell by opening the two top flat plates. Pressure gauges are attached with the membrane housing to maintain the gas flow rate at different transmembrane pressures. The gas is purged into the membrane unit through a pipe from the top plate. The gas is then permeated through the outer wall of the tubular cell unit, connected to the bubble flow meter for measuring the gas flow rate at different transmembrane pressures. A rubber balloon is attached at one end of the bubble flow meter, which contains soap solution. The gas permeation experiment is conducted at room temperature ($28 \pm 2°C$).

3.3.3 Microscopic Techniques

Microscopic techniques are considered as standard methods in ceramic membrane characterization. These techniques deliver information about the morphology of membrane surfaces, thickness of cross sections, and surface roughness as well as grain size.

3.3.3.1 Scanning Electron Microscopy

A SEM is a form of electron microscope that produces images of a sample by scanning it with a focused beam of electrons rather than light. The signals that originate from electron–sample interactions mainly provide qualitative information about the sample, including external surface texture, morphology (pore size and pore size distribution) using ImageJ software [5], and chemical composition with the help of energy-dispersive x-ray spectroscopy (EDS). In most of the applications, data are collected over a particular zone of the surface of the sample depending upon the image quality, and an image is generated that shows spatial differences in these properties. In conventional SEM techniques, magnification can range from 20× to approximately 30,000× with a spatial resolution of 50 to 100 nm. This technique can produce good image quality of materials only from 1 cm to approximately 5 µm.

Working principle: Figure 3.6 illustrates the working principle of the SEM. The details of this principle are as follows:

a. Electrons are generated at the top of the microscope by a metallic filament (A) referred to as the electron gun.

b. The emitted electrons are then formed into a beam (B) and accelerated down the column toward the specimen.

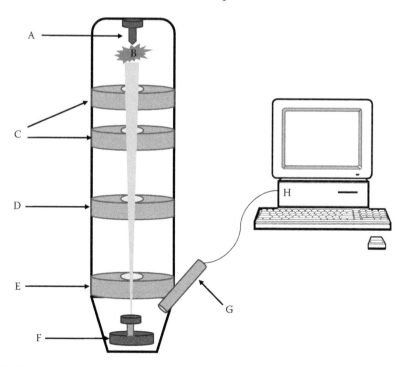

FIGURE 3.6
A graphical presentation of the working principle of SEM.

c. The beam is further focused and directed by electromagnetic lenses (C–E) as it moves down the column.

d. When the beam reaches the specimen in the objective chamber, electrons are freed from the surface of the specimen (referred to as secondary electrons).

e. These electrons are captured by a detector (G) that intensifies the signal and sends it to a monitor (H).

f. The electron beam scans back and forth across the sample, building up an image by calculating the number of electrons emitted from each spot over the sample.

g. This entire procedure takes place inside a vacuum cylinder. The vacuum makes sure that the electron beam interacts with the sample rather than the air; otherwise, a blurry image will be produced due to the formation of moisture over the sample surface.

h. Samples used for SEM analysis must be able to withstand a vacuum and need to be conductive.

i. Ceramic samples that are nonconductive in nature can be coated with an ultrathin coating of electrically conducting material and deposited on the sample either by low-vacuum sputter coating or by high-vacuum evaporation. Conductive materials in current use for specimen coating include gold, gold/palladium alloy, platinum, osmium, iridium, tungsten, chromium, and graphite.

3.3.3.2 Field Emission Scanning Electron Microscopy

FESEM provides information related to surface topography and elemental composition (using an energy dispersive x-ray [EDX]) of ceramic membranes. It can operate at 10× to 300,000× magnification, with virtually unlimited depth of field. FESEM produces flawless, less electrostatically distorted images with spatial resolution in nanometer ranges (three to six times better compared to SEM). The advantages of FESEM include the ability to inspect smaller area contamination spots at electron accelerating voltages compatible with EDS, reduced penetration of low-kinetic-energy electron probes closer to the immediate surface of the material, and high-quality, low-voltage images with negligible electrical charging of samples. Nowadays, in-lens FESEM is also used for ultrahigh magnification imaging.

The *working principle* is as follows:

a. Electrons, known as primary electrons, are liberated from a field emission source and accelerated in a high electrical field gradient.

b. These primary electrons are then focused within the high vacuum column and deflected by electromagnetic lenses (condenser lens, scan coils, stigmator coils, and objective lens) to produce a narrow scan beam that bombards the object.

c. As a result, secondary electrons are emitted from each spot of the object.

d. The angle and velocity of these secondary electrons relate to the surface structure of the object.

e. A detector is introduced to capture the secondary electrons, which produces an electronic signal.

f. This signal is amplified and transformed to a video-scan image that can be seen on a monitor. The data obtained from the monitor can be saved and processed further.

Figure 3.7 demonstrates the working principle of FESEM and each segment of electromagnetic lenses. Electromagnetic lenses consist of four different segments: condenser lens, scan coils, stigmator coils, and objective lens. The current in the condenser lens defines the diameter of the beam. A low current indicates a small diameter, whereas a higher current signifies a larger beam. A narrow beam produces better resolution but is also responsible for inferior signal-to-noise ratio. A reverse situation ensues when the beam has a large diameter. The function of scan coils is to deflect the electron beam over the object in a zig-zag arrangement. The objective lens is the lowest lens in the column. The objective focuses the electron beam on the object. The shortest working distance provides the smallest beam diameter and the best resolution, but also the poorest depth of field that indicates which range in the vertical direction in the object can still be visualized sharply. The stigmator coils are applied to correct irregularities in the x- and y-deflection of the beam and thus to attain a perfectly round-shaped beam. These beams produce good-quality images, whereas an ellipsoidal beam is responsible for blurred and stretched images.

3.3.3.3 Atomic Force Microscopy

AFM uses a very high resolution type of scanning probe microscope. It is mainly used to verify the roughness of the surface of the ceramic membrane. The schematic of the working principle of AFM is shown in Figure 3.8. The AFM technique has several advantages over SEM. It can produce images of materials ≤1 nm, which is much smaller than SEM (usually around 100 nm). The AFM technique provides a three-dimensional surface image, unlike the other conventional microscopic method. In spite of that, no sample preparation is required for this method as it can work smoothly in ambient air or even a liquid environment. But the problem associated with this method is the area of the image (scanning area: approx. 150 × 150 μm). Another disadvantage is the quality of the image (high resolution, less quality).

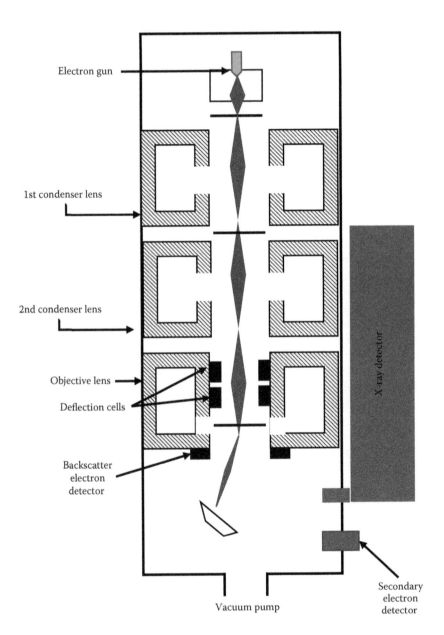

FIGURE 3.7
Schematic diagram of the working principle of FESEM.

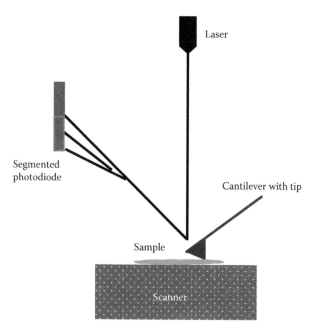

FIGURE 3.8
A graphic view of the operating principle of AFM.

The *operating principle* of AFM can be divided into different sections:

- Surface sensing
    - A cantilever, made of silicon or silicon nitride and with a very sharp tip radius (<5 nm), scans a sample surface.
    - The cantilever deflects toward the surface due to attractive force between the surface and the tip when the tip is brought into close range of the sample surface, according to Hook's law [12].
- Detection
    - A laser beam is used to measure cantilever deflections reflected from the flat top of the cantilever.
    - A position-sensitive photo diode (PSPD) is used to track these deviations.
- Imaging
    - Traditionally, the sample is mounted on a piezoelectric tube. The tube can move in x- and y-directions for scanning the sample and the z-direction for maintaining a constant force.
    - AFM images of the topography of a sample surface are taken by scanning the cantilever over an area of interest.

- The deflection of the cantilever can be realized by the raised (brighter zone) and lowered (dark zone) features on the sample surface, which is monitored by the PSPD.
- The height of the tip above the surface can be controlled by using a feedback loop. It is recommended to maintain constant laser position to obtain an accurate plot of surface topography.

3.3.3.4 Transmission Electron Microscopy

With a very high resolution (about 1 nm), TEM is another technique used to observe morphology of membrane surfaces. In this microscope, an electron beam from an electron gun is transmitted through an ultrathin section of the sample ceramic membrane and the image is magnified by the electromagnetic fields. Basically, it is difficult to observe particulars of internal structures of ceramic membranes because of small cracks in samples caused by the cutting procedure during sample preparation.

For TEM, the ceramic specimen has to be dry and extremely thin; it is introduced into the microscope at a point above the electromagnetic lens (Figure 3.9).

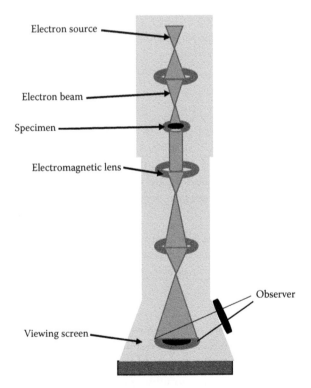

FIGURE 3.9
A schematic of the operating principle of TEM.

TEM offers very high magnification and can also provide information on element and compound structure. But this is very expensive and requires special housing and maintenance. Sample preparation for this method is also laborious and analysis requires special training.

3.4 Measurement of Mechanical Strength

Mechanical properties are important in ceramic membrane fabrication. They include the many properties used to describe the strength of ceramic membranes, such as flexural strength, fracture toughness, and hardness. Ceramic materials are usually ionic or covalent bonded materials, and they can be crystalline or amorphous. Ceramic materials shows poor toughness and tensile strength as these materials are held together by a weak bond, causing fracture before any plastic deformation. Moreover, these materials are likely to be porous and imperfect (a few cases) and act as stress concentrators, decreasing the toughness further and reducing the tensile strength. Therefore, it is necessary to study the mechanical behavior of the ceramic membrane to understand its morphological as well as mechanical properties.

3.4.1 Flexural Strength

Flexural strength is a material property, defined as the stress in a material just before it yields in a flexure test. It is also known as modulus of rupture, bend strength, or fracture strength. The transverse bending test is most commonly used: A specimen with either a circular or rectangular cross section is bent until fracture or yielding takes place, using either a three-point or four-point flexural bend test technique. The three-point bend test is more suitable for ceramics compared to the four-point bend test. As in the former case, the specimen requires minimum size, which increases the measured value of flexural strength.

Three-point bend test (ASTM D790): This test provides values for the modulus of elasticity in bending, flexural stress, flexural strain, and the flexural stress–strain response of the material. The advantage of a three-point flexural test is the ease of the sample preparation and testing. But, this method also has some disadvantages, such as that obtained results are sensitive to ceramic sample and loading geometry and strain rate.

In this test, a rectangular, bar-shaped specimen has to be prepared by maintaining thickness/length ratio as 1:16 and is placed on two parallel supporting pins, as shown in Figure 3.10. The force is applied in the middle of the specimen by means of a loading pin. The supporting and loading pins are mounted in a way to allow their free rotation about the axis parallel to the pin and specimen axis.

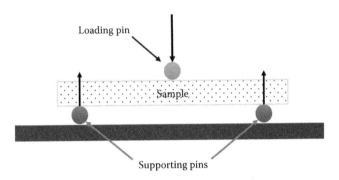

FIGURE 3.10
Schematic of three-point bend test for determining flexural strength.

3.4.2 Fracture Toughness

Fracture toughness is an important property of advanced ceramics and is one measure of brittleness—less popular in the field of ceramic membrane fabrication. The aim of a fracture toughness test is to measure the resistance of a material to the presence of a flaw in terms of the load required to cause a brittle or ductile crack extension in a standard specimen containing a fatigue precrack. There are some standard tests (ASTM C1421-16) available to determine fracture toughness of ceramics [13,14]. Charpy and Izod tests are well-known techniques for measuring fracture toughness. The Charpy v-notch test (ASTM A370) is a standardized high strain-rate test that determines the amount of energy absorbed by a material during fracture. This absorbed energy is a measure of a given material's notch toughness and acts as a tool to study temperature-dependent ductile–brittle transition. The standard specimen size utilized for this test is $10 \times 10 \times 55$ mm. It is widely applied in industry, since it is easy to prepare and conduct and the results can be obtained quickly and cheaply. The disadvantage of this method is that the obtained results are only comparative. The Izod test is the same as the Charpy test; the only difference is in the specimen arrangement ($63.5 \times 12.7 \times 3.2$ mm).

3.4.3 Hardness

Hardness is a measure of a material's resistance to localized plastic deformation. Hardness tests are simple, inexpensive and nondestructive. In this technique, no special specimen need be prepared, and the testing apparatus is relatively inexpensive. Other mechanical properties often may be estimated from hardness data, such as tensile strength. Rockwell (ASTM E-18), Brinell, Knoop (HK), and Vickers (HV) microindentation methods are some commonly used tests for the determination of hardness of ceramic membranes.

3.5 Determination of Chemical Stability

3.5.1 Acid–Alkali Test

Chemical stability of ceramic membranes after long-term exposure to a harsh environment is usually verified by immersing them into the acid (concentrated HCl, preferably in pH range of 1 to 2) and alkali (NaOH, preferably in pH range of 12 to 14) solution for 7 days. Firstly, weight of the specimen is measured in dry conditions and then dipped into the solution of acid and alkali for 1 week at atmospheric conditions. Secondly, the specimen is removed from the sample and then weighed and the porosity estimated following the volumetric porosity determination technique. EDX analysis of the membranes before and after the corrosion test needs to be performed to verify the change in elemental composition.

3.6 Characterization of Membrane Surface Charge

3.6.1 Contact Angle

Ceramic membranes in an aqueous environment have an attractive or repulsive response to water. The material composition of the membrane and its corresponding surface chemistry control the interaction with water, thus affecting its wettability. The wettability of a material is a surface property that yields a distinctive value for each and every material. The surface tension value of a material can be used to determine wettability of a material by specific liquids. The surface tension for the solid material can be calculated through the measurement of the contact angle between a solid surface and a droplet of liquid on the surface. Ceramic membranes are classified in two different categories depending upon the surface–water interaction: hydrophobic (or hydrophobicity) or hydrophilic (or hydrophilicity). Hydrophobic ("water hating") materials have little or no tendency to adsorb water on their surface, whereas hydrophilic ("water loving") materials adsorb water readily. Hydrophobicity and hydrophilicity of ceramic materials are commonly characterized by the contact angle measurement technique (Figure 3.11). The contact angle is the angle, conventionally measured through the liquid, where a liquid–vapor interface meets a solid surface. The shape of a liquid–vapor interface (Table 3.1) is determined by the Young–Laplace equation. The theoretical description of contact arises from the consideration of a thermodynamic equilibrium between the three phases: the liquid phase (L), the solid phase (S), and the gas or vapor phase (V). If the solid–vapor interfacial energy is denoted by γ_{SV}, the solid–liquid interfacial energy by γ_{SL}, and the liquid–vapor interfacial energy (i.e., the surface tension) by γ_{LV},

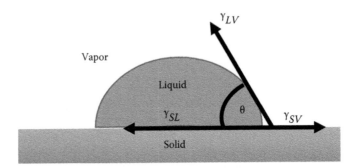

FIGURE 3.11
Thermodynamic equilibrium between the vapor, liquid, and the solid phases.

TABLE 3.1

Wettability of Ceramic Membranes Based on Contact Angle

Contact Angle	Wettability	Hydrophilicity
$\theta = 180°$	Spontaneous	None
$180° > \theta > 90°$	Good	Poor
$\theta = 90°$	Moderate	Moderate
$90° > \theta > 0°$	Poor	Good
$\theta = 0°$	None	Outstanding

then the equilibrium contact angle θ_C is determined from these quantities by Young's equation:

$$\gamma_{SV} = \gamma_{SL} + \gamma_{LV} \cos \theta \qquad (3.14)$$

3.6.2 Zeta Potential

Zeta potential is one of the most appropriate parameters controlling the rheological behavior of ceramic suspensions. The control of the stability of a ceramic dispersion can avoid particle aggregation occurring. The measurement of zeta potential and rheological properties can be used to optimize dispersion conditions to give a defect-free product.

A conventional ceramic process, such as slip casting, requires stable, well-dispersed colloidal suspensions to attain optimal casting behavior and green body properties. Colloidal particles are nearly always electrically charged. If the charge is adequately high, the electrical repulsive forces will stop the particles from agglomerating, and stable suspensions with low relative viscosity can be prepared.

The particle charge can be operated and organized by regulating the suspension pH and by using appropriate dispersants, such as polyelectrolytes. However, optimizing the formulation of these materials requires a suitable method for measuring the particle charge, known as the zeta potential.

The charge on colloidal particles can arise from a number of different mechanisms, including dissociation of acidic or basic groups on the particle surface, or adsorption of a charged species from solution. The particle charge is balanced by an equal and opposite charge carried by ions in the surrounding liquid. These counter ions tend to cluster around the particles in diffuse clouds. This arrangement of particle surface charge surrounded by a diffuse cloud of countercharge is called the electrical double layer (Figure 3.12).

The electrical potential drops off exponentially with distance from the particle and reaches a uniform value in the solvent outside the diffuse double layer. The zeta potential is the voltage difference between a plane that is a short distance from the particle surface and the solvent beyond the double layer.

When two particles come so close that their double layers overlap, they repel each other. The strength of this electrostatic force depends on the zeta potential, which is a key indicator of the stability of colloidal dispersions. The magnitude of the zeta potential indicates the degree of electrostatic repulsion between adjacent, similarly charged particles in a dispersion. For

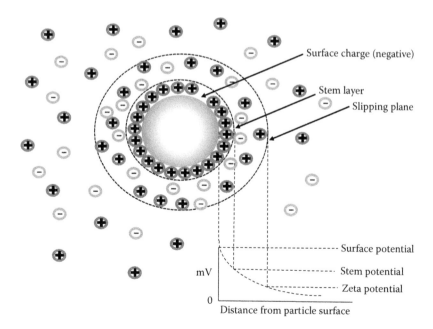

FIGURE 3.12
Schematic diagram of the fundamentals of zeta potential.

molecules and particles that are small enough, a high zeta potential will confer stability (i.e., the solution or dispersion will resist aggregation). When the potential is small, attractive forces may exceed this repulsion and the dispersion may break and flocculate. So, colloids with high zeta potential (negative or positive) are electrically stabilized, while colloids with low zeta potentials tend to coagulate or flocculate as outlined in Table 3.2.

A high zeta potential will prevent particle–particle agglomeration and keep the dispersion uniform and free flowing. Therefore, the goal in most formulations is to maximize the zeta potential. This is particularly important when trying to produce high-strength ceramic materials.

The zeta potential depends on the surface charge density and the double layer thickness. The surface charge density, in turn, depends on the concentration of potential-determining ions in the solvent—ions that have a particular affinity for the surface. In many ceramic systems, the H+ ion is potential determining, so the zeta potential depends on pH.

The zeta potential is positive for low pH values and negative for high pH values. The pH at which the zeta is zero is the isoelectric point (IEP) of the colloid. The IEP is a property of the particle surface. For alumina, the IEP is usually around 9.5. Thus, alumina slurries are usually stable below about pH 8.

Zeta potential is not measurable directly, but it can be calculated using theoretical models and an experimentally determined electrophoretic mobility or dynamic electrophoretic mobility. Electrokinetic and electroacoustic phenomena are the usual sources of data for calculation of zeta potential.

Electrokinetic phenomena arise when two phases move with respect to each other with an electric double layer at the interface. Electrophoresis is used for estimating zeta potential of particulates, whereas streaming potential/current is used for porous bodies and flat surfaces. In practice, the zeta potential of dispersion is measured by applying an electric field across the dispersion. Particles within the dispersion with a zeta potential will migrate toward the electrode of opposite charge with a velocity proportional to the magnitude of the zeta potential.

Electrophoresis is the movement of a charged surface stationary liquid applied electric field. Electrophoretic velocity is proportional to electrophoretic

TABLE 3.2

Stability Behavior of the Colloid Based on Zeta Potential Value

Zeta Potential [mV]	Stability Behavior of the Colloid
From 0 to ±5	Rapid coagulation or flocculation
From ±10 to ±30	Incipient instability
From ±30 to ±40	Moderate stability
From ±40 to ±60	Good stability
More than ±61	Excellent stability

mobility, which is the measurable parameter. There are several theories that link electrophoretic mobility with zeta potential. Electro-osmosis is the counterpart of electrophoresis, that is, the movement of charged particles in a fluid under the influence of an electric field.

Two electroacoustic effects are widely used for characterizing zeta potential: colloid vibration current and electric sonic amplitude. There are commercially available instruments, known as electrophoretic light scattering (ELS), for measuring dynamic electrophoretic mobility. This mobility is often transformed to zeta potential to enable comparison of materials under different experimental conditions.

Electroacoustic techniques have the benefit of being able to perform measurements in intact samples, without dilution. Calculation of zeta potential from dynamic electrophoretic mobility requires information on the densities for particles and liquid. In addition, for larger particles exceeding roughly 300 nm in size, information on the particle size is required as well.

3.7 Other Techniques

3.7.1 XRD

X-ray diffraction (XRD), an essential and unique technique for identification of phase, structure, and chemical composition of crystalline ceramic specimens, is described in detail in numerous texts [15–18]. It provides qualitative and quantitative information about any changes in crystalline structures of membrane raw materials before and after the sintering process.

The diffraction of x-rays by crystal was first identified by Bragg in 1912. Bragg explained the condition for constructive interference, which gives rise to intense diffraction maxima; this is known as Bragg's law (Figure 3.13):

$$2d \sin \theta = n\lambda_1 \tag{3.15}$$

where
d is the spacing between the lattice planes of the crystal
θ is the incidence angle
n is the order of diffraction and is an integer
λ_1 is the wavelength of the x-rays

The following is the derivation of Bragg's law:

$$\Delta ABZ, \; n\lambda_1 = AB + BC$$

Based on geometry, $AB = BC$. Therefore, $n\lambda_1 = 2AB$.

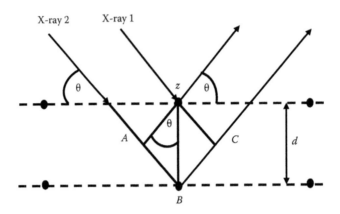

FIGURE 3.13
Diffraction of x-rays by a crystal.

According to geometrical definition, $\sin\theta = AB/d$. Then, $AB = d\sin\theta$. Now, we can write, $2d\sin\theta = n\lambda_1$.

The conditions for constructive interference are (1) the atomic spacing in the solid must be comparable with the wavelength of the x-rays, and (2) the scattering centers must be spatially distributed in an ordered way. When the angle of incidence (θ) does not satisfy Equation 3.15, constructive interference fails and destructive interference occurs.

In the past, the amount of work necessary for x-ray analysis of structure and composition was time consuming and prolonged. But recently, it has been reduced as much of the data analysis is now usually performed by software and computerized programs. X-ray analysis can be categorized into two segments: (1) structural analysis and (2) compositional analysis. Structural analysis includes the measurement of the lattice parameters of the crystal and the crystal structure by using a single crystal method as well as a powder diffraction technique (the sum of the patterns of the individual materials). Compositional analysis is based on a unique x-ray diffraction pattern for each crystalline material. Hence, chemical identity of a compound can be assumed by matching the x-ray diffraction pattern of the unknown material and an authentic sample. The x-ray diffraction patterns for several thousands of materials are basically provided by the ICDD-JCPDS (International Center for Diffraction Data-Joint Committee on Powder Diffraction Standards).

3.7.2 SAXS

SAXS (small-angle x-ray scattering) is an analytical method to determine the information related to averaged particle sizes, shapes, geometry and bulk morphology at a molecular level for a wide variety of samples, such as metals, ceramics, polymers, biological molecules, etc., in the forms of solids, liquids, liquid dispersions, films, gels, ground powders, and molecules in gaseous phase. When such a sample is exposed by x-rays, the sample scatters the radiation differently depending on its constituents to create a contrast that helps to draw conclusions about the particle structure and its arrangement in the system. Apart from this basic information, much other information can be obtained from SAXS—namely, characteristic distances of partially ordered materials, radius of gyration (R_g) to calculate particle dimensions, molecular weight of particles, structures, polydispersity analysis, molecular interactions, their orientations, degree of crystallinity, etc.

The technique presently is used in quality control and varied areas of research (structural biology, chemistry, physics, and engineering). The basic components of a SAXS instrument are source, collimation system, sample holder, beam stopper, and a detection system. The source generates the x-ray of desired wave lengths; the collimator makes the beam narrow and zero-angle.

The beam stopper has a dual role: It prevents the high-intensity rays from directly reaching the detector, and thus prevents destruction, and also allows only scattered signals to reach the detector. The scattered beam coming from a sample is then detected on the other side using a two-dimensional area detector. Similarly, the surface of the particles can be measured selectively when the x-rays hit a flat surface almost parallel to the surface. This gives information about relative positional order on the surface or within the surface layer.

SAXS is used for characterization of structures at nanometer-length scale. The characterization of structure by SAXS involves studies of the following parameters:

- Average size in the nanometer range of 0.1 to 200 nm
- Shape of structure in the above-size range
- Internal architecture of the nanostructure (e.g., core-shell, lamella, single body, etc.)
- Measuring the ordered structures such as fractals, core sizes, shell thickness, lamellar spacing, etc.
- Measurement of crystalline domains and porous system (closed and open) at nanoscales
- Measurement of nanosize distribution and surface-to-volume ratio of nanoparticles
- Measurement of protein folding kinetics in native media conditions
- Measurement of a polymer blend's interaction and interface architecture
- Measurement of nanocomposites' internal organization and structure–property correlation
- Measurement of macromolecular molar mass

Other operational features are described in the following text.

3.7.2.1 Protein Structure and Function

SAXS and proteins in solution SAXS allow obtaining the three-dimensional structure of proteins and protein assemblies in solution, which is essential for a better understanding of their biological function. The knowledge of the three-dimensional structure of proteins is the key to understanding their biological function. X-ray structure analysis of protein crystals is a well-established method to obtain this structural information. However, the method has the disadvantage that proteins must be crystallized, which is often a serious problem, and that the protein structure in the crystal does not necessarily reflect its structure under biological conditions. The great advantage of SAXS for protein research is that investigations can be performed in solution under biological conditions and that structural changes can be studied upon changing the external conditions. Due to its high x-ray flux in the sample, the measuring times with the SAXS technique are low. This is especially important for the investigation of biological/protein samples, where long radiation exposure time may lead to radiation damage.

3.7.2.2 Drug Delivery Systems

Dispersions with liquid–crystalline nanosized substructures are of great importance in different scientific and industrial applications. These systems exhibit nanostructures with high interfacial area, low viscosity, and the capability to solubilize various molecules. They are important as membrane mimetic matrices, as vehicles for active ingredients (e.g., vitamins and enzymes), and as unique microenvironments for the controlled release of additives (e.g., drugs). Transfer of material occurs when droplets with different ingredients are mixed. This happens without fusion or dissolution of droplets and even the droplet size does not change significantly. Since the internal structure depends on the composition, one can follow the material transfer (i.e., the uptake or the release of ingredients) by monitoring structure changes with time-resolved SAXS. This technique fulfills the main requirements for successful time-resolved SAXS experiments: high intensity and high resolution resulting in short measurement times at good quality data. The uptake and release kinetics of any molecules can be studied with the SAXS technique in the same way, if there is a dependence of the internal structure on the concentration of the molecules under investigation. Droplets by themselves might be stable for months, if not years, but they can easily exchange their contents. So for describing the stability of a nanostructured emulsion, determination of the particle size has to be complemented by studying the internal structure.

3.7.2.3 Pharmaceutical Formulations

A microemulsion is a thermodynamically stable dispersion of one liquid phase into another, stabilized by an interfacial film of surfactant.

Microemulsions can be used as a medium for phase transfer catalysis in formulations of lubricating fluids in pharmaceutical or cosmetic applications. The structural information gathered by means of the SAXS experiments serves to improve formulations based on microemulsions. Small- and wide-angle x-ray scattering (SWAXS) is a powerful tool to characterize samples containing nanosized lamellar structures. These structures play an important role in pharmaceutical applications for the development and characterization of new drug formulations. SAXS determines the self-assembled (nanosized) structures of such materials (i.e., lamella repeat distance and long range order). Wide-angle x-ray scattering (WAXS) determines the crystallinity at the atomic level and therefore delivers the characteristic fingerprint of these materials. The SWAXS technique precisely and simultaneously measures the small- and wide-angle scattering of pharmaceutical excipients. It enables determination of both the lamellar nanostructure and atomic crystallinity in less than a minute.

3.7.2.4 Structure and Internal Core-Shell Structure of Nanoparticles

The SAXS technique can be used to characterize nanostructured lamellae present in a surfactant sample. The internal structure (core shell) of the lamellae can be precisely determined (i.e., the thickness of the core and the shell, lamella repeat distance, and the long-range order). These structures play an important role in several surfactant applications like detergents and drug carriers as well as in biomembranes. The SAXS technique can be used to investigate the formation of silica nanoparticles stabilized by organic molecules. SAXS in combination with small-angle neutron scattering studies allows determination of the size and the core-shell structure of the nanoparticles, as well as the impact of concentration variations on their formation. This may help to understand a variety of processes from the synthesis of zeolites and ordered mesoporous silica to, for example, biomineralization or water-rock kinetics in geosciences.

3.7.2.5 Polymer Nanocomposites

Inorganic nanoparticles (e.g., soot, silica particles, layered silicates, etc.) are used as fillers to enhance the performance of polymeric materials. The SAXS technique can easily and quickly monitor the morphological state of layered silicates in polymeric matrices. This technique offers the possibility to monitor the small-angle and wide-angle regions at the same time. The degree of intercalation and exfoliation can be estimated, and the processing conditions and the material used can be optimized according to the results. In this application note, we report on the influence of particle concentration and temperature on the exfoliation of natural clay particles in water and surface-modified clay particles in organic solvent. We can use the SAXS technique for these studies. SAXS allows us to study the scattering of platelet-like clays and offers a convenient and sensitive method to study exfoliation as a function of processing conditions.

3.7.2.6 Colloids and Microemulsions

The particles in emulsions (e.g., oil droplets in water) and suspensions (e.g., enzymes, colloidal inorganic precipitates, etc.) are usually stabilized with surfactants or surface charges against coalescence or clustering (aggregation) processes. In order to determine the size distribution of dispersed particles, a number of different methods are available which differ greatly in speed and resolution capabilities. Static light or x-ray scattering techniques employ a moderate resolution of 20% peak-to-peak separation with medium time consumption (minutes to hours). Scattering techniques can be effectively used to determine the formation of aggregates. This is due to the fact that the intensity of scattered x-rays is roughly proportional to the sixth power of the particle size. This effectively means that a few aggregates can be detected at a very early stage due to their strong influence on the scattering signal. Very high quality data are vital to the success of particle sizing with scattering techniques. The SAXS technique has proved to deliver excellent data that are demonstrated to be well suited for particle sizing.

3.7.2.7 Thin Films and GISAXS

Thin films of mesoporous materials are currently studied intensively for applications in future energy conversion or storage systems. Because of the porous nature of these films, high interfacial surface areas for electron transport can be obtained. Control of these interfacial structures is essential for the performance of such systems. It is also important to study the dependence of structure and possible structural (phase) changes on the temperature. Grazing incidence small-angle scattering (GISAXS) is a highly sensitive and time-saving analysis method for the characterization of such thin films. In GISAXS, the x-ray beam is directed onto the sample at a very shallow angle ($i < 1°$). By changing this incident angle, the penetration depth of the x-rays into the material can be altered. Thus, structures close to the very surface of the film and buried structures can be discriminated. The heating module for the GISAXS stage enables the user to perform such temperature-dependent GISAXS studies at temperatures of up to +500°C in vacuum, air, or inert gas.

3.7.2.8 Surfactant Systems

The structure of inhomogeneous (core-shell) nanoparticles can be studied with the SAXS technique. The internal structure of sodium dodecyl sulfate (SDS) micelles in water was determined by calculating the radial electron density profile. For example, SDS is a highly effective anionic surfactant used in many hygiene and cleaning products or drug carrier systems. In aqueous solution, SDS molecules self-assemble and form micelles: SAXS allows one to obtain the shape and size of such nanosized micelles and—due to its sensitivity to electron-density differences—to determine the internal (core-shell) structure. This is of great importance for understanding and controlling the role of surfactants in different materials (e.g., the stability of emulsions or the release rate of the active ingredient in drug carrier systems).

3.7.2.9 Characterization of Physical Properties of Liquid Crystalline Compounds

Phase-change materials absorb, store, and release heat when they undergo a phase transition, such as from solid to liquid. Paraffins, salt hydrates, and fatty acids can be used as phase-change materials. Application fields are building materials (e.g., drywalls or solar heat storage), packaging, and clothes. A typical phase-change material is encapsulated paraffin wax, which is used to increase the thermal mass of the material in which it is embedded. Depending on the intended application, the melting range of the wax is chosen. This report compares DSC measurements of encapsulated paraffin wax and combined SWAXS measurements performed in the same temperature range. The SAXS technique allows monitoring the changes in a material upon changing the temperature. The temperature-controlled sample holder allows rapid temperature changes and, by comparing with DSC, shows that the temperature adjustment is precise. Since the SAXS technique offers the possibility to monitor the small- and wide-angle regions at the same time, and measurements take only 1 min, it allows for measurements that are not easily feasible with classical SAXS and/or WAXS equipment.

3.7.2.10 Functionalized Quantum Dots

Quantum dots (QDs) are semiconducting materials whose properties and characteristics are strongly related to their size. QDs are used in various applications such as LED technology and biological imaging techniques. In comparison to, for example, traditional organic fluorescent dyes, the color of quantum dots is much more stable and much brighter. The precise determination of size and size distribution of quantum dots is very important as the size determines properties such as fluorescence, conductivity, and other properties. For example, after excitation, small quantum dots emit light with a smaller wavelength (blue shift) compared to larger quantum dots (red shift). Ideally, all particles in a sample should have a narrow size distribution, ensuring a precise (i.e., monochromatic) wavelength of the emitted light. Therefore, SAXS is a helpful method to control the properties of quantum dots because it quickly determines the particle size and produces representative results.

3.7.2.11 Porous Nanostructures and Catalysts

Zeolites are of great importance in different industrial applications. They are widely used as catalysts for important reactions like cracking and hydrocarbon synthesis. Reactions can take place within the zeolite pores, which allows a high degree of product control. The porous structure of zeolites also acts as a "sieve" for molecules with certain dimensions. This can be utilized (e.g., for gas separation). Furthermore, cations within the zeolite framework can readily exchange with other cations in aqueous media. This ion exchange capability makes zeolites applicable (e.g., as water softening devices in detergents and soaps). Zeolites are synthesized from aqueous alkaline gels or solutions containing sources of silica or alumina, and cations. Variation of the inorganic base and the use of organic templating cations (e.g., tetra-alkylammonium [TAA] cations) results in a range of products with specific properties. Even small changes in the mixture concentration, pressure, or thermal history have an impact on the properties of the final materials. Understanding many of the observed phenomena still requires a deeper knowledge about the processes occurring on the molecular level during synthesis. One of them is the formation of nanoparticles prior to and during zeolite crystal growth. In this report the formation of silica nanoparticles in TAA hydroxide solutions is studied using the small-angle x-ray scattering technique. A mesoporous material was studied with the SAXS technique. Important structural parameters like pore size, degree of polydispersity, total volume fraction of the pores, and arrangement of the pores were obtained. Porous materials can have an enormous internal surface. They are used as catalysts, as molecular sieves, as adsorbents, and in many more applications. Their ability to selectively adsorb molecules of a certain size and shape is closely related to their internal pore size and geometry. SAXS is an ideal tool for the characterization of the structure and the pore size of such materials (i.e., the particle diameter D and the mean pore diameter d, as well as the typical distance between the pores R, can be determined).

3.7.3 XPS

The instrument XPS (x-ray photoelectron spectroscopy) is the most widely used surface analysis technique. In this method, the sample is irradiated with monoenergetic x-rays, causing electrons to be emitted from the sample surface. An electron energy analyzer determines the binding energy of the electrons. From the binding energy and intensity of an electron peak, the elemental identity, chemical state, and quantity of an element are determined. The information XPS provides about surface layers or thin film structures is of value in many applications, including polymer surface modification, catalysis, corrosion, adhesion, semiconductor and dielectric materials, magnetic media, and thin film coatings. Using this instrument, additional activities, such as imaging XPS, UV analysis, Auger electron analysis, etc., can also be done.

The instrument can be used for surface analysis of any material, such as polymers, magnetically permeable materials, ceramics, metals, semiconductors, etc.—to name a few that can be easily analyzed:

1. Materials science and engineering—understanding the surface chemistry of different materials from various engineering processes, redox reactions, kinetics, and mechanisms
2. Photochemistry and photocatalysis—photon illumination of the sample surface at the analysis position through an optical fiber from an external UV-visible light source enables understanding of surface reactions, modifications, and decompositions
3. Chemical mapping of biological and environmental materials—studying the metal distribution in mineral and biological surfaces as well as transport and migration in groundwater
4. Nanoscience and technology—revealing the surface and interface interactions of nanoparticles with organic ligands, polymers, and biological films
5. Solar cells based on semiconductors and polymers—providing molecular distribution, the nature of binding and charge transfer, and the resulting electronic structure and energy level alignments

Typical specifications: dual anode Mg/Al x-ray source, multichannel detection detector with >10 Mcps count rate and data acquisition software, 50 mm z-shift, port aligner, water cooling system, control software, multielement small spot lens (<60 μm): three different magnifications, acceptance angles and exit apertures, 180° double focusing geometry, set of lead-glass windows, optical sample alignment, fine focus ion source, multimetal analysis chamber ~300 mm ID, ~25 mm diameter sample holder, preparation facility, imaging package.

General applications of XPS include

- Surface analysis of organic and inorganic materials
- Determining composition and chemical state information from surfaces
- Depth profiling for thin film composition
- Thin film oxide thickness measurements
- Examining polymer functionality before and after processing
- Bonding and adhesion issues
- Obtaining depth profiles of thin film stacks (both conducting and nonconducting) for matrix level constituents
- Identifying stains and discolorations
- Characterizing cleaning processes

3.7.3.1 Applications in Metals

XPS is a fundamental characterization tool for investigating a wide range of surface problems on metal and oxide surfaces. With its surface selectivity and quantifiable data, XPS is the ideal tool for measuring composition and thickness of protective oxide films on metals. Finally, XPS can detect any contaminants that may have been introduced during the manufacturing process. Surface contamination of finished components happens due to manufacturing or process failures, which cause performance issues as well as cross-contamination, corrosion, and electrical contact problems. Many of these surface contamination issues are difficult to detect during or after production. XPS is commonly used to identify surface contaminants and to pinpoint manufacturing or process failures that contribute to material surface impurity. XPS combines quantitative elemental and chemical information with extreme surface sensitivity. Since XPS provides information on oxides and chemical states of materials, it is highly useful in the analysis of aluminum surfaces and their interface to other materials and environments. Analysis of nonperforming parts after manufacture can also be carried out to determine failure.

3.7.3.2 For Catalyst and Nanocomposite Membrane Materials

Surface analysis is paramount to understanding the reactivity, selectivity, and catalytic ability of substances. XPS analysis is quite helpful in understanding the surface charges and binding energies for the different preparation methods and compositions used in catalyst preparation. The changes during the operation in catalyst structure due to different types of fuels and products formed during the operation and the catalyst stability studies can be easily done by XPS. In the development of membrane electrode assembly (MEA), where catalyst distribution is very important over the large area, can be investigated in XPS. Along with this, it can be helpful in identification of surface or near surface species and nanoparticle distribution.

3.7.3.3 Pharmaceutical Applications

XPS analysis is an important analytical tool for quantitatively probing the chemistry of the outer few nanometers of pharmaceutical materials. XPS is a quantitative, surface-sensitive spectroscopy capable of probing the local chemical bonding of inorganic and organic materials. XPS is used to quantify the amount of various chemicals present on the outer surface of the prepared polymers. This analysis can also be performed for aggregates as well as for individual particles to assess surface chemical uniformity. Understanding the surface concentration of various components in pharmaceutical powders is critical for controlling performance parameters, such as solubility, dissolution, stability, flow ability, agglomeration, and crystallinity. The interactions among the particles can also be analyzed by XPS analysis, which is highly recommended for pharmaceutical applications.

3.7.3.4 Applications in Surface Modification

XPS is a fundamental characterization tool for investigating a wide range of surface problems on metal and oxide surfaces. With its surface selectivity and quantifiable data, XPS is the ideal tool for measuring composition and thickness of protective oxide films on metals. In addition, sputter profiling the steel surface provides a full understanding of the surface composition and chemistry, which may help to diagnose failures in the passivation process. Finally, XPS can detect any contaminants that may have been introduced during the manufacturing process. The information XPS provides about surface layers or thin film structures is important for many industrial and research applications where surface or thin film composition plays a critical role in performance including: nanomaterials, photovoltaics, catalysis, corrosion, adhesion, electronic devices and packaging, magnetic media, display technology, surface treatments, and thin film coatings used for numerous applications.

3.8 Summary

After fabrication of a ceramic membrane, it is necessary to characterize the membrane to understand its thermal, morphological, mechanical, and chemical behavior; this is possible only if the membrane passes through different characterization techniques. Thus, different characterization procedures have been detailed in this chapter.

The overall outline of this chapter is given as follows:

- How do we measure thermal stability of a ceramic membrane or how do we know about the thermal stability of the raw materials used in fabrication of a ceramic membrane?
  - TGA provides information about the temperature regimes where the raw materials, especially the pore-formers and additives, are decomposing.
- Determination of morphological properties:
  - Physical methods
    - Mercury porosimetry—pore size and pore size distribution
    - Perporometry—active pores and their size distribution
    - Thermoporometry—pore size (both open and closed)

- – Gas adsorption/desorption isotherms—pore size and its distribution
- – Archimedes' principle—porosity
- – Bubble point—size of largest pore
- – Measurement of solute rejection—detection of complete solute rejection
- • Permeation methods
  - – Liquid permeation—average pore radius
  - – Gas permeation—pore radius and effective porosity
- • Microscopic techniques
  - – SEM
  - – FESEM
  - – AFM
  - – TEM
- • How to know mechanical strength of a ceramic membrane
  - • Flexural strength—three-point bend test (ASTM D790)
  - • Fracture toughness—Charpy and Izod tests
  - • Hardness—Rockwell (ASTM E-18), Brinell, Knoop (HK), and Vickers (HV) microindentation methods
- • How to determine chemical stability of a ceramic membrane
  - • Acid–alkali test
- • How to understand membrane surface charge
  - • Contact angle
  - • Zeta potential
- • Other advanced techniques to investigate membrane surfaces
  - • XRD
  - • SAXS
  - • XPS

References

1. K. Li, *Ceramic membranes for separation and reaction*, John Wiley & Sons Ltd., Chichester, England, 2007, ISBN 978-0-470-01440-0.
2. C. Eyraud, M. Betemps, J.F. Quinson, F. Chatelut, M. Brun, B. Rasneur, Determination of the pore-size distribution of an ultrafilter by gas liquid perporometry measurement—Comparison between flow porometry and

condensate equilibrium porometry. *Bulletin de la Societe Chimique de France partie—I Physicochimie des systemes liquides electrochimie catalyse genie chimique,* 9–10 (1984) 1237–1244.

3. G.Z. Cao, J. Meijerink, H.W. Brinkman, A.J. Burgrraaf, Perporometry study on the size distribution of active pores in porous ceramic membranes. *Journal of Membrane Science,* 83 (2) (1993) 221–235.

4. M. Brun, A. Lallemand, J.F. Quinson, C. Eyraud, New method for simultaneous determination of size and shape of pores—Thermoporometry. *Thermochimica Acta,* 21 (1) (1977) 59–88.

5. S. Bose, C. Das, Preparation and characterization of low cost tubular ceramic support membranes using sawdust as a pore-former. *Materials Letters,* 110 (2013) 152–155.

6. S. Bose, C. Das, Role of binder and preparation pressure in tubular ceramic membrane processing: Design and optimization study using response surface methodology (RSM). *Industrial & Engineering Chemistry Research,* 53 (2014) 12319–12329.

7. S. Bose, C. Das, Sawdust: From wood waste to pore-former in the fabrication of ceramic membrane. *Ceramics International,* 41 (3) part A (2015) 4070–4079.

8. K. Nath, *Membrane separation processes,* PHI Pvt. Ltd., New Delhi, India, 2008, ISBN-978-81-203-3532-5.

9. W.S.W. Ho, K.K. Sirkar (eds.), *Membrane handbook,* Van Nostrand Reinhold, New York, 1992.

10. P. Silva, A.G. Livingston, Effect of solute concentration and mass transfer limitations on transport in organic solvent nanofiltration—Partially rejected solute. *Journal of Membrane Science,* 280 (1–2) (2006) 889–898.

11. J. Marchese, C.L. Pagliero, Characterization of assymetric polysulphone membranes for gas separation. *Gas Separation and Purification,* 5 (4) (1991) 215–221.

12. A.T. Hubbard, *The handbook of surface imaging and visualization,* CRC Press, Boca Raton, FL, 1995.

13. G.D. Quinn, J. Salem, I. Baron, K. Cho, M. Foley, H. Fang, Fracture toughness of advanced ceramics at room temperature. *Journal of Research of the National Institute of Standards and Technology,* 97 (1992) 579–607.

14. J.J. Swab, J. Tice, A.A. Wereszczak, R.H. Kraft, Fracture toughness of advanced structural ceramics: Applying ASTM C1421. *Journal of American Ceramic Society,* 98 (2) (2014) 607–615.

15. B.D. Cullity, *Elements of x-ray diffraction,* 2nd ed., Addison-Wesley, Reading, MA, 1978.

16. R. Jenkins, R.L. Snyder, *Introduction to x-ray powder diffractometry,* John Wiley & Sons, New York, 1996.

17. H.P. Klug, L.F. Alexander, *X-ray diffraction procedures for polycrystalline and amorphous materials,* John Wiley & Sons, New York, 1974.

18. G.H. Stout, L.H. Jensen, *X-ray structure determination,* John Wiley & Sons, New York, 1989.

4

Ceramic Membrane Cleaning Methods

4.1 Introduction

In recent years, ceramic membrane has experienced a significant growth in the field of novel separation processes, and it has become one of the major technologies for the recovery and/or removal of different air pollutants such as carbon dioxide (CO_2), volatile organic compounds (VOCs), elemental sulfur (S_X), oxides of nitrogen (NO_X), acid gases, and so on. In addition, it has been widely used in chemical as well as in biochemical applications. Despite the popularity of ceramic membrane over every aspect of the separation process, it has major drawbacks that need to be overcome. Specifically, fouling due to chemical affinity between solutes and membrane material at the membrane surface or inside the pores, pore blockage, and formation of gel are the most vital problems since the economics of the process and the performance of the membrane are highly influenced by the rate of fouling. Fouling is really a huge problem causing declination of flux, enhancement of consumption of energy, and deterioration of permeate quality, thus reducing the performance of the membrane and increasing the cost of the process. Hence, cleaning the membrane is necessary to overcome this problem for enhancing the lifetime of the membrane and productivity. Other possible ways to control fouling are feed pretreatment and operation conditions [1–3]. Membrane cleaning methods can be classified into three different techniques: physical, chemical, and physicochemical, discussed briefly in this chapter.

4.2 Physical Cleaning Techniques

Physical cleaning methods use mechanical forces to dislodge and remove foulants from the membrane surface. Physical methods include sponge ball cleaning, forward and reverse flushing, backwashing with deionized water, air bubbling or air sparging, and back permeation by gas (preferably CO_2) [4–7]. In addition, ultrasonic [8–10], spark discharge [11], electrical field [12], and magnetic field [13] are other recently developed physical cleaning methods that are commonly used for membrane cleaning purposes.

4.2.1 Sponge Ball Cleaning

In this technique, sponge balls (especially polyurethane) are inserted into the membrane modules for a few seconds to scrub the foulant from the membrane surface [5]. They are normally used for cleaning large tubular membranes for wastewater and industrial process water [14].

4.2.2 Flushing

Flushing can be either forward or reverse in nature (Figure 4.1). In the forward flushing method, permeate water is pumped at high cross-flow velocity through the feed side in order to remove foulants from the membrane surface [5].

Due to high cross-flow velocity and the subsequent turbulence, the absorbed particles inside the membrane wall and pores are released and discharged (e.g., removal of colloidal matter). The reverse flushing method demonstrates

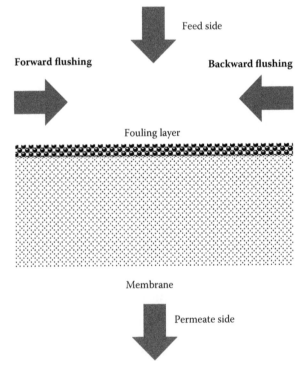

FIGURE 4.1
Forward and reverse flushing.

an alternative direction of the permeate flush for a few seconds in the forward direction and a few seconds in the reverse direction.

4.2.3 Backwashing

In this filtration process, permeate is flushed in a reverse way through the membrane to the concentrate side. In the case of porous ceramic membranes, the pores are flushed inside out due to higher membrane pressure on the permeate side than the pressure within the membranes when backward flush is applied (Figure 4.2). Backwashing is commonly used in reverse osmosis membranes either by reducing operating pressure below the osmotic pressure of the feed solution or by increasing the permeate pressure [15].

4.2.4 Air Sparging

This method produces a two-phase flow to eliminate external fouling and thus diminishes the cake layer deposited on the membrane surface. Air sparging can be applied either during the course of filtration to reduce fouling deposition [16–18] or periodically to remove already formed deposits. The types of gases used for the sparging of gas are water/CO_2 mixture and water/N_2 mixture. Air sparging is normally applied in microfiltration (MF), ultrafiltration (UF), and nanofiltration (NF) membranes with

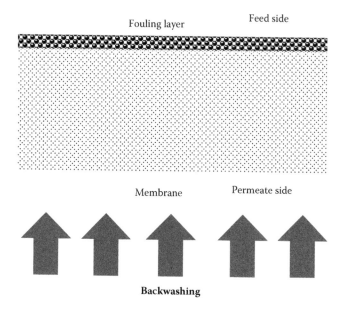

FIGURE 4.2
Flow direction for cleaning of membrane by backwashing.

tubular, flat sheet, and, to some extent, hollow fiber and spiral wound modules. The purpose of applying air is to achieve an enhanced flux with high separation efficiency in MF and UF systems. This is possible due to the presence of air bubbles, which intensify turbulence in the feed side of the membrane, thus increasing permeate flux as well as solute separation efficiency [16,19].

4.3 Chemical Cleaning Technique

Chemical cleaning is the most widely used membrane cleaning method, especially in ceramic membranes. In this cleaning process, a cleaning agent (often a mixture of compounds) plays the major role, so the choice of cleaning agent is critical. Generally, selection of the suitable cleaning agent is done on the basis of membrane material and type of fouling (colloidal, organic, metal oxides, silica, carbonate scales, sulfate scales, etc.). These cleaning agents dissolve most of the deposited materials on the surface and remove them without damaging the membrane surface. Normally, acidic cleaning agents, such as nitric, phosphoric, hydrochloric, sulfuric, and citric acids, are used to remove precipitated salts from the membrane surface, whereas alkaline cleaning agents are suitable for the removal of organic foulants. However, the major drawback of strong acids is their great impact on the pH of the solutions. Too low a pH can affect the reliability of membranes. In this respect, weak acids are usually preferred as cleaning agents in membrane plants. Phosphoric acid (H_3PO_4) and some organic acids, such as citric acid, are good buffers in maintaining pH during cleaning and therefore much less corrosive. Other kinds of chemical cleaning agents are surfactants (nonionic, anionic, cationic, and zwitterionic)—for example, cetyltrimethyl-ammonium bromide (CTAB), sodium dodecyl sulfate (SDS), etc.—sequestration agents (EDTA), dispersants or deflocculants, metal chelating agents, and enzymes [3]. A typical cleaning cycle generally consists of the following stages:

- Removal of product
- Rinsing with water
- Repetitive cleaning steps
- Rinsing with water again

Some physical transformations, such as melting, thermal and mechanical stress, wetting, soaking, and swelling, among others, and chemical reactions, such as peptization, hydrolysis, solubilization, dispersion, chelation, sequestering, and suspending, take place when the cleaning agents come into contact with the fouled layer [20]. The other significant factors regarding chemical cleaning are temperature, chemical concentration, pH, pressure and flow, and time [21,22].

4.4. Physicochemical Cleaning Technique

Many situations arise where physicochemical interactions occur between solution species and the membrane material. Physicochemical properties, including charge effects and hydrophobicity, cause a concentration profile and deposition of feed solution over the membrane surface. The charges on a membrane are strongly dependent upon the membrane material, the pH, and the ionic strength of the feed solution. To minimize the concentration profile, two other cleaning techniques are gaining attention; one is the physicochemical technique and the other is the electrochemical method. The physicochemical cleaning methods use physical cleaning methods with the addition of chemical agents to improve cleaning efficiency. The applications usually involve forward flushing with permeate between cleanings when more than one chemical cleaning is used, but not simultaneous use of physical and chemical cleaning actions [5].

Electrochemical methods permit cleaning membranes at low cross-flow velocity due to the use of potential application between electrodes. This method has not been implemented extensively in any large plant yet. It is appropriate in cleaning metallic microfilters when they have been fouled with albumin and phosphate.

4.5 Other Techniques

Many novel and nonconventional techniques, such as application of ultrasound, electric field, and magnetic field to the membrane surface, have been developed to overcome fouling without reducing the lifetime and efficiency of the membrane.

4.5.1 Ultrasound-Assisted Cleaning

Ultrasound is produced at frequencies over 20000 Hz by electromechanical transducers based on q-factor piezoelectric effect. The basic physical phenomenon behind the effect of ultrasound is bubble cavitation. This acts as a successive ultrasound wave through the liquid medium in a series of alternate compression and expansion cycles. Cavitation mainly starts between the aforesaid ranges of frequency and promotes formation, growth, and implosive collapse of bubbles in the liquid that generates high-velocity and shock waves at the time of contact with a solid surface, which has substantial mechanical and chemical effects [23]. This phenomenon causes reduction in fouling deposited over the membrane surface. High-velocity fluid can decrease the thickness of the boundary layer and diffusional resistance and therefore enhances the rate of mass transfer. The main advantages of ultrasound cleaning are the following [24]:

- Continuous cleaning of membrane at the time of operation
- No use of harmful chemicals during cleaning
- Production of hydrogen peroxide (H_2O_2) and hydroxyl free radical ($\cdot OH$) by ultrasound, which can be used as a disinfectant for drinking water

The major problem associated with this process is high energy consumption by the ultrasound transducers. The important variables in this process are the intensity of the ultrasound field, the solute concentration, and the irradiation time.

Recently, application of ultrasound in chemical cleaning of ceramic membranes for different membrane processes has been studied extensively [25–28].

4.5.2 Electric Field-Assisted Cleaning

The electric field is considered an alternative way to minimize membrane fouling. This phenomenon is based on electrokinetics, electrophoresis, and electro-osmosis. Electrokinetics has a positive effect in the filtration process that contributes as an additional driving force without increasing any shear force. Electrophoresis is associated with the movement of the solids or charged species (e.g., proteins) [25]. Electro-osmosis is nothing but a movement of the fluid through a porous membrane (permeation).

Basically, the particles are attracted by an electrophoretic effect on the charged molecules due to the electric field and are lifted and carried over from the membrane surface, thus reducing the concentration polarization layer and increasing the flux.

Another significant effect due to the application of an electric field is electrolysis (ion migration), which can affect filtration performance and the fouling process. The strength of the electric field applied over the membrane surface depends on the conductivity of the feed, electrode material and its location, and the interaction between the membrane and the substances being used. Zeta potential value and charge of the species present in the feed should be investigated for further development of the technology.

The main advantage of this cleaning technique is that the electric field can be applied in the work cycle without any interval such as the ultrasound cleaning technique discussed before [29]. But the cost of the electrode makes this technique unfavorable compared to the other techniques. Another problem related to the process is that the area of the electrode should not interfere with the feed and permeate flows [30].

4.5.3 Magnetic Field-Assisted Cleaning

Application of the magnetic field for cleaning the membrane is mostly used in the treatment of water from different industrial processes such as pigment, brightener filler, adsorbent where there are possibility of membrane clogging, and pore blocking due to the scaling of calcium carbonate ($CaCO_3$). This technique is economical, easy to operate, and environmentally benign. Magnetic fields are applied in the feed flow, in the storage or reactor tank, and even over the pipe that takes the flow into the membrane module. Devices for generating magnetic fields may be electromagnets or permanent magnets.

4.5.4 Pulsatile Flow

A pulsed flow is imposed on a stationary flow with frequencies of 1 Hz. Thus, the shear stresses applying to the particles increase and the cleaning result improves. Pulsatile flow is nothing but a particle oscillatory motion on surfaces caused by mechanical vibrations of the surfaces. The oscillation depends on the frequency and direction of the external force. An oscillating force can induce weakening and breaking of the bonds between particle and surface. In recent studies, the influence of a laminar pulsed flow with low frequencies and high-velocity amplitudes during pipe cleaning has been examined [31]. In addition, the investigations presented that the laminar pulsed flow can improve the cleaning result for the cleaning of surfaces. In contrast, others have investigated the effect of pulsating turbulent flows on wall shear stress components in cylindrical pipes [32,33]. They showed that the pulse flow causes an increase in the velocity gradient at the wall of the pipe, which is responsible for better cleaning.

4.6 Summary

In spite of the reputation of ceramic membranes for every feature of the separation process, they have major drawbacks, such as fouling at the membrane surface or inside the pores, pore blockage and formation of a gel that cause reduction in the performance of the membrane due to declination of flux, high energy consumption, and reduced permeate quality. Hence, cleaning the membrane is essential to enhance its life span. This chapter has given information about different cleaning techniques that can be summarized as

- Membrane cleaning methods—physical, chemical, and physico-chemical techniques
- Physical cleaning methods—using mechanical forces to remove foulants from the membrane surface
  - Sponge ball cleaning
  - Forward and reverse flushing
  - Backwashing with deionized water
  - Air bubbling or air sparging
  - Back permeation by gas
- Chemical cleaning methods—CTAB, SDS, EDTA, and phosphoric acid; different steps of chemical cleaning methods
  - Removal of product
  - Rinsing with water
  - Repetitive cleaning steps
  - Rinsing with water
- Physicochemical cleaning technique—use physical cleaning methods with the addition of chemical agents to improve cleaning efficiency
- Other techniques—ultrasound-assisted cleaning, electric field-assisted cleaning, magnetic field-assisted cleaning, and pulsatile flow

References

1. J.J. Sadhwani, J.M. Veza, Cleaning tests for seawater reverse osmosis membranes. *Desalination*, 139 (1–3) (2001) 177–182.
2. M.F.A. Goosen, S.S. Sablani, H. Al-Hinai, S. Al-Obeidani, R. Al-Belushi, D. Jackson, Fouling of reverse osmosis and ultrafiltration membranes: A critical review. *Separation Science and Technology*, 39 (10) (2004) 2261–2297.
3. T. Mohammadi, S.S. Madaeni, M.K. Moghadam, Investigation of membrane fouling. *Desalination*, 153 (1–3) (2002) 155–160.
4. A. Al-Amoudi, W.L. Lovitt, Fouling strategies and the cleaning system of NF membranes and factors affecting cleaning efficiency. *Journal of Membrane Science*, 303 (1–2) (2007) 4–28.
5. S. Ebrahim, Cleaning and regeneration of membranes in desalination and wastewater applications: State of the art. *Desalination*, 96 (1–3) (1994) 225–238.

6. A. Fouladitajar, F.Z. Ashtiani, H. Rezaei, A. Haghmoradi, A. Kargari, Gas sparging to enhance permeate flux and reduce fouling resistances in cross flow microfiltration. *Journal of Industrial and Engineering Chemistry*, 20 (2) (2014) 624–632.
7. A. Ghadimkhani, W. Zhang, T. Marhaba, Ceramic membrane defouling (cleaning) by air nano bubbles. *Chemosphere*, 146 (2016) 379–384.
8. S. Popovic, M. Djuric, S. Milanovic, M.N. Tekic, N. Lukic, Application of an ultrasound field in chemical cleaning of ceramic tubular membrane fouled with whey proteins. *Journal of Food Engineering*, 101 (2010) 296–302.
9. M.O. Lamminen, H.W. Walker, L.K. Weavers, Mechanisms and factors influencing the ultrasonic cleaning of particle-fouled ceramic membranes. *Journal of Membrane Science*, 237 (2004) 213–223.
10. E. Alventosa-deLara, S. Barredo-Damas, M.I. Alcaina-Miranda, M.I. Iborra-Clar, Study and optimization of the ultrasound-enhanced cleaning of an ultrafiltration ceramic membrane through a combined experimental-statistical approach. *Ultrasonics Sonochemistry*, 21 (2014) 1222–1234.
11. H.-S. Kim, K. Wright, D.J. Cho, Y.I. Cho, Self-cleaning filtration with spark discharge in produced water. *International Journal of Heat and Mass Transfer*, 88 (2015) 527–537.
12. X. Chen, H. Deng, Effects of electric fields on the removal of ultraviolet filters by ultrafiltration membranes. *Journal of Colloid and Interface Science*, 393 (2013) 429–437.
13. M. Gryta, The influence of magnetic water treatment on $CaCO_3$ scale formation in membrane distillation process. *Separation and Purification Technology*, 80 (2) (2011) 293–299.
14. C. Psoch, S. Schiewer, Direct filtration of natural and simulated river water with air sparging and sponge ball for fouling control. *Desalination*, 197 (1–3) (2006) 190–204.
15. A. Sagiv, R. Semiat, Parameters affecting backwash variables of RO membranes. *Desalination*, 261 (3) (2010) 347–353.
16. Z. Cui, T. Taha, Enhancement of ultrafiltration using gas sparging: A comparison of different membrane modules. *Journal of Chemical Technology and Biotechnology*, 78 (2–3) (2003) 249–253.
17. M. Mercier, C. Fonade, C. Lafforgue-Delorme, How slug flow can enhance the ultrafiltration flux in mineral tubular membranes. *Journal of Membrane Science*, 128 (1) (1997) 103–113.
18. C. Cabassud, S. Laborie, L. Durand-Bourlier, J.M. Lainé, Air sparging in ultrafiltration hollow fibers: Relationship between flux enhancement, cake characteristics and hydrodynamic parameters. *Journal of Membrane Science*, 181 (1) (2001) 57–69.
19. G. Ducom, C. Cabassud, Possible effects of air sparging for nanofiltration of salted solutions. *Desalination*, 156 (1–3) (2003) 267–274.
20. G. Tragardh, Membrane cleaning. *Desalination*, 71 (3) (1989) 325–335.
21. R. Liikanen, J. Yli-Kuivila, R. Laukkanen, Efficiency of various chemical cleanings for nanofiltration membrane fouled by conventionally treated surface water. *Journal of Membrane Science*, 195 (2) (2002) 265–276.
22. J.P. Chen, S.L. Kim, Y.P. Ting, Optimization of membrane physical and chemical cleaning by a statistically designed approach. *Journal of Membrane Science*, 219 (1–2) (2003) 27–45.

23. J. Li, R.D. Sanderson, E.P. Jacobs, Non-invasive visualization of the fouling of microfiltration membranes by ultrasonic time-domain reflectometry. *Journal of Membrane Science*, 201 (1–2) (2002) 17–29.
24. J.Y. Lu, X. Du, G. Lipscomb, Cleaning membranes with focused ultrasound beams for drinking water treatment, *Proceedings of IEEE International Ultrasonics Symposium Proceedings*, 1195–1198, Rome, September, 2009.
25. A. Saxena, B.P. Tripathi, M. Kumar, V.K. Shahi, Membrane-based techniques for the separation and purification of proteins: An overview. *Advances in Colloid and Interface Science*, 145 (1–2) (2009) 1–22.
26. S. Popovic, M. Djuric, S. Milanovic, M.N. Tekic, N. Lukic, Application of an ultrasound field in chemical cleaning of ceramic tubular membrane fouled with whey proteins. *Journal of Food Engineering*, 101 (2010) 296–302.
27. M.O. Lamminen, H.W. Walker, L.K. Weavers, Mechanisms and factors influencing the ultrasonic cleaning of particle-fouled ceramic membranes. *Journal of Membrane Science*, 237 (2004) 213–223.
28. E.A-. deLara, S.B-. Damas, M.I.A-. Miranda, M.I.I-. Clar, Study and optimization of the ultrasound-enhanced cleaning of an ultrafiltration ceramic membrane through a combined experimental statistical approach. *Ultrasonics Sonochemistry*, 21 (2014) 1222–1234.
29. E. Iritani, Y. Mukai, Y. Kiyotomo, Effects of electric field on dynamic behaviours of dead-end inclined and downward ultrafiltration of protein solutions. *Journal of Membrane Science*, 164 (1–2) (2000) 51–57.
30. S.N. Jagannadh, H.S. Muralidhara, Electrokinetics methods to control membrane fouling. *Industrial Engineering & Chemistry Research*, 35 (4) (2006) 1133–1140.
31. C.R. Gillham, P.J. Fryer, A.P.M. Hasting, D.I. Wilson, Enhanced cleaning of whey protein soils using pulsed flows. *Journal of Food Engineering*, 46 (3) (2000) 199–209.
32. W. Blel, C. Le Gentil-Lelièvre, T. Bénézech, J. Legrand, P. Legentilhomme, Application of turbulent pulsating flows to the bacterial removal during a cleaning-in-place procedure. Part 1: Experimental analysis of wall shear stress in a cylindrical pipe. *Journal of Food Engineering*, 90 (4) (2009) 422–432.
33. W. Blel, P. Legentilhomme, T. Bénézech, J. Legrand, C. Le Gentil-Lelièvre, Application of turbulent pulsating flows to the bacterial removal during a cleaning-in-place procedure. Part 2: Effects on cleaning efficiency. *Journal of Food Engineering*, 90 (4) (2009) 433–440.

5

Introduction to Membrane Reactors and Membrane Contactors

5.1 Membrane Reactors

5.1.1 Classification of MRs Based on Their Configuration

Catalytic reactors based on ceramic membrane are classified into three categories: (a) inert membrane reactors (IMRs), used to add or remove certain species from the reactor without direct participation of membrane in the reaction; (b) catalytic membrane reactors (CMRs), when a catalyst becomes catalytically active during preparation by addition of catalyst precursors; and (c) a combination of CMRs and IMRs, catalytically active both inside and outside the membrane. These categories are explained by different configurations labeled as configurations A, B, C, D, E, F, G, and H, as shown in Figure 5.1. Inert membrane reactors are explained by configuration A, used to add or remove particular species from a reactor without direct participation of the membrane in the reaction. The removal of at least one reaction product involves equilibrium displacement and offers higher reaction yield—for example, removal of hydrogen in dehydrogenation reactions is the most widely used application for this type of configuration. This configuration has also been fit to other processes such as decomposition of H_2S and H_2O and production of synthesis gas [1,2].

Configuration A provides both selective (e.g., Pd- or Ag-based alloys on ceramic substrates) and preferential (removal of product silica, alumina, titania, glass, zeolite membranes, etc.) separation, as confirmed by several scientists. In this case, the membrane does not participate in the reaction directly, but it is used to add or remove certain species from the reactor. The most widely used application involves equilibrium displacement by removal of at least one reaction product [3,4].

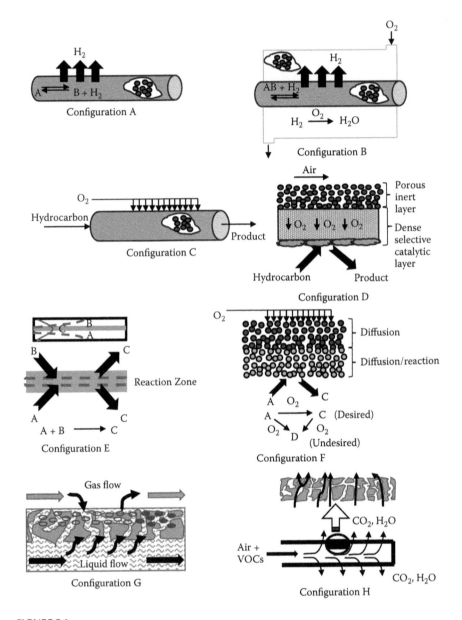

FIGURE 5.1
Membrane reactors based on the principle of configurations.

Configuration B offers higher yield through reaction coupling. In this case, on both sides of the membrane, complementary processes are run that use either the permeated species (chemical coupling, e.g., dehydrogenation/ hydrogenation, or dehydrogenation/combustion reactions), or the heat generated in the reaction (thermal coupling, exothermic/endothermic processes).

The reactions often use different catalysts, which would be packed on opposite sides of the membrane tube [5,6].

Configuration C describes the distribution of a reactant to a fixed bed of catalyst packed in the opposite side of the membrane. Inert membrane reactors (IMRs) of this configuration (meso- or microporous membranes) have already been used successfully as oxygen distributors in methane oxidative coupling [7,8] and the production of olefins and oxygenates [9,10] from the oxidation of alkanes. The membrane used for the distribution of oxygen in oxidation processes is safer as it reduces formation of hot spots and runaway reactions and gives greater selectivity through control of the concentration of the selected species along the reactor with respect to the conventional feed arrangements. Porous membranes with a dense layer can also be applied for reactant distribution but have some difficulties in attaining high permeation fluxes related to oxidation reactions.

Reactant distribution can also be achieved using porous membranes with a thin but dense permselective layer. In the case of oxidation reactions, this would have the important advantage of using air instead of oxygen in the oxygen supply side. However, few results have been reported to date, which is mainly due to the difficulties in attaining sufficiently high permeation fluxes (which is usually achieved by reducing the thickness of the dense layer), while at the same time maintaining the membrane properties during prolonged exposure at operating conditions. The only clear advantage with respect to reactant distribution using porous membranes would be available from configuration D, where, at least in principle, the oxygen species transported through the membrane could react before recombination and desorption take place. This would completely avoid the presence of gas phase oxygen and could certainly represent a valuable alternative, provided that a membrane with sufficiently high reaction selectivity and permeability to oxygen can be developed. In this case, the reaction on one side of the membrane acts as an efficient oxygen sink, resulting in enhanced oxygen transport across the membrane.

Configuration E defines the separation of reactants on both sides of a porous catalytic membrane in which the reaction takes place in a small zone or a plane inside the porous structure, as the reaction rate is higher than the mass transport. This avoids the slip of reactants to the opposite side and also helps to reduce unwanted side reactions. By changing the reactant concentrations outside the membrane, the position of the reaction plane can be shifted to a new location where transport rates to the reaction zone are again matched by the reaction stoichiometry. This gives a lower residence time in the reaction zone and also reduces further reaction of partial oxidation products [11]. The same principle has been demonstrated with experiments and model calculation by CO [12] and H_2S oxidation [13]. The performance of a membrane reactor with separate feeding of reactant for the catalytic combustion of methane based on this concept has been reported [14].

Configuration F is a modified version of configuration C. The objective of this configuration is to minimize the concentration of oxygen in the reacting environment and to deliver a sharp distribution of the active component across the entire membrane [15–17]. In this case, the goal is the same (reduce the concentration of oxygen in the reacting environment), although the oxygen's partial pressure is now lowered by feeding it through a diffusion layer of sufficient resistance. Oxygen diffusion can take place by itself or in the presence of a stagnant fluid filling the pores of the membrane. This can be achieved by feeding an inert species at approximately the same partial pressure to both sides of the membrane.

The diffusion zone is followed by a catalytic layer, where the reaction of oxygen and the reactant permeated from the opposite side takes place.

The purpose of using configuration G is to improve the contact in gas–liquid–solid systems by providing a distinct contact zone, as shown in Figure 5.1. This configuration has the potential to overcome the difficulty of catalyst recovery that appears in slurry reactors. The concept has already been made popular in the case of hydrogenation reactions over Pt/Al_2O_3 catalysts [18,19].

Configuration H delivers the same idea as configuration G. The last two configurations aim to improve contact efficiency as well as conversion by decreasing mass transfer resistance. This approach has been employed to prepare catalytically modified fly ash filters for alcohol dehydration and for reduction of nitrogen oxides with NH_3 [20,21].

5.2 The Catalytic Membrane Reactor and Its Novel Applications

5.2.1 Synergistic Effect of Separation and Reaction

The conventional packed-bed reactor has the limitation of the reaction equilibrium. In a packed-bed reactor the reactants are premixed and brought into contact with the reaction products, inducing the uncontrollable reaction between products and intermediate or reactant to the undesirable one. The synergistic effect of separation and reaction occurs simultaneously within a single unit and is particularly apparent for reactions limited by thermodynamic equilibrium considerations—for example, catalytic hydrocarbon dehydrogenation or esterification reactions [22].

The membrane functions as the separator of the desired product from the mixture stream, encouraging the reaction to proceed on the right side of reaction and producing more products—in other words, shifting the equilibrium of the reaction "to the right" (according to Le Chatelier's principle) for higher conversion. Meanwhile, this will enhance the selectivity and yield of the product by preventing further conversion of the desired product to the undesired one for certain reactions.

Some of the porous inert membranes, however, are used as the distributor of the reactants in order to elongate the reaction site along the membrane, providing more reaction site but mitigating the hot spot of reaction zone as compared to the catalyst bed in the packed-bed reactor.

5.2.2 Membrane Functions in CMRs and Applications

The basic functions of porous ceramic membranes in membrane reactors are divided into three primary categories, as shown in Figure 5.2 [23]:

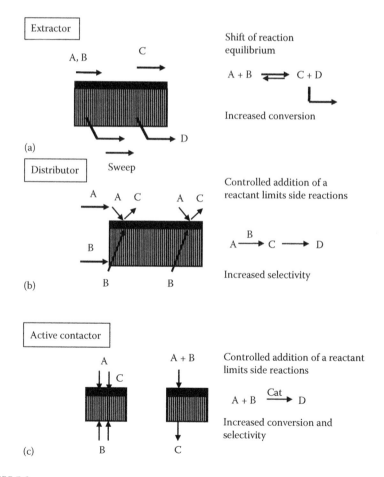

FIGURE 5.2
The three main membrane functions in a membrane reactor. (a) Extractor provides high conversion by shifting of reaction equilibrium, (b) distributor offers high selectivity by the controlled addition of reactant(s), and (c) active contactor provides both high conversion and selectivity.

1. **An extractor:** the removal of product(s) increases the reaction conversion by shifting the reaction equilibrium.
2. **A distributor:** the controlled addition of reactant(s) limits side reactions and increases the selectivity.
3. **An active contactor:** the controlled diffusion of reactants to the catalyst can lead to a catalytic reaction zone and offers higher conversion and selectivity.

The MR extractor mode has been successfully investigated to increase the conversion of a number of equilibrium-limited hydrogen-producing reactions such as alkane dehydrogenation, by selectively extracting the hydrogen produced [24], the water gas shift, the steam re-forming of methane, the decomposition of H_2S and HI, etc. The two important factors controlling the efficiency of the process are H_2 permselectivity of the membrane and its permeability. Except for H_2 removal, some decomposition reactions in which O_2 is removed have also been studied in extractor mode [25]. Reactive separations for light alkane dehydrogenation reactions to produce olefins have also been effectively executed using inorganic porous membranes.

The propane dehydrogenation reaction has been studied in a packed-bed membrane reactor (PBMR) using a microporous Si/Al_2O_3 membrane made by chemical vapor deposition or chemical vapor infiltration (CVD/CVI) with a H_2 permeance of around 1.4×10^{-9} mol.m^{-2}.Pa^{-1}.s^{-1}, and a H_2/C_3H_8 permselectivity in the range of 70–90 at 500°C. The authors reported that conversion was twice as high as the equilibrium value in the PBMR at 500°C [26,27].

Parallel reactions such as oxydehydrogenation (ODH) of hydrocarbons and oxidative coupling of methane are typical examples of the distributor mode MR. Membranes are generally used to control the supply of O_2 in a fixed bed of catalyst in order to bypass the flammability area and to separate the alkane from O_2 in oxidative coupling of methane. Membranes are applied in MRs to optimize the O_2 profile concentration along the reactor and to maximize the selectivity in the desired oxygenate product. This concept also plays a beneficial role for the high temperature in exothermic reactions [28]. The first study of ODH of hydrocarbons using porous membranes was in the CMR configuration [29].

In the active contactor mode, the membrane acts as a diffusion barrier and is not permselective, but catalytically active. The concept is used with a forced flow mode or with an opposing reactant mode; that is, the transportation of one reactant through the membrane is in the opposite direction of another reactant. The forced flow contactor mode is studied for enzyme-catalyzed reactions [24], total oxidation of volatile organic compounds [30], oxidative coupling of methane [31], selective hydrogenations [32], and photocatalytic oxidation of carbonyl compounds and bioaerosols [33]. The selectivity of alkene hydrogenation triphasic reactions can also be enhanced when both the alkene and hydrogen are forced to pass through a microporous catalytic membrane [34].

The opposing reactant contactor mode applies to both equilibrium and irreversible reactions [35]. In this reactor, the different reactants are fed separately at the two sides of a catalytically active membrane. In such a case, a small reaction zone forms in the membrane in which reactants are in stoichiometric ratio. In this arrangement, a permselective membrane is not essential; the membrane only supplies the reaction zone. The partial pressure difference is the force that helps the reactants to diffuse toward each other in the catalytic zone.

The aim behind this concept is to attain full conversion of the pollutant and to avoid its slippage out of the reactor. If the reaction kinetics is faster than the mass transport in the membrane, a reaction zone is formed that prevents the side reactions. This concept has been demonstrated for the selective catalytic reduction (SCR) of nitric oxide with ammonia to nitrogen and water [36]. The Claus reaction between hydrogen sulfide and sulfur dioxide in a CMR using this concept were studied earlier [11]. Hydrogen sulfide (H_2S) and sulfur dioxide (SO_2) have been fed on the opposite sides of the membrane and the reaction takes place within the membrane itself. It has been reported that, using this kind of system, undesired side reactions can be avoided by properly adjusting the operating conditions. Triphasic (gas/liquid/solid) reactions (e.g., olefin hydrogenation) can also be improved by using this concept, which is limited by the diffusion of the volatile reactant [18,37]. The other applications using this concept are in the area of oxidation of carbon monoxide, catalytic combustion of propane, oxidative dehydrogenation of propane, catalytic combustion of methane, catalytic combustion of natural gas, etc.

5.2.3 Opposite Flow Mode Catalytic Membrane Reactors

The basic principles and applications of the opposite reactant mode CMR concept along with various alternative membrane reactor concepts are briefly discussed next. The concept is shown in Figure 5.3. The literature dealing with catalytic membrane reactors in opposite-flow mode is limited. However, the concept is highly appreciable, performing several catalytic reactions in the gas phase.

Catalytic membrane reactors with separate feed of reactants result in excellent mass transfer between fluid and membrane and avoid slip (the reaction takes place entirely inside the membrane). Using this kind of membrane reactor can minimize thermal runaways and can assist in avoiding undesired side reactions by properly adjusting the operating conditions. These types of reactors are easily controlled and exclusively applied for combustion processes as no explosive mixture will build up and safety will increase. Claus reactions are studied to realize the effect of reversibility of the reaction and pressure difference on the flux of the gaseous mixture.

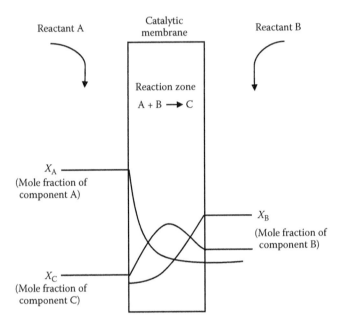

FIGURE 5.3
Membrane reactor with separate feed of reactants.

5.2.4 Hydrogen Sulfide (H₂S) Laden Gas Treatment

The first application of a catalytic membrane reactor was reported using a simplified model with separate feed of reactant [38]. By means of mathematical modeling of molecular diffusion and viscous flow of an instantaneous and reversible Claus reaction [39] inside the membrane, catalytic membranes were prepared with α-Al$_2$O$_3$ with a mean pore diameter of 350 nm, a porosity of 41%, and a thickness of 4.5 mm and impregnated with γ-Al$_2$O$_3$. In this study, a mathematical model was developed to understand the effect of reversibility of the reaction, and pressure difference on the flux of the gaseous mixture in the membrane reactor consisted of two well mixed chambers.

The influence of reversibility was investigated by means of a simulation carried out at temperatures of 200°C and 300°C in absence of a pressure difference. The simulated results inferred that reversibility of the reaction mainly arises for a lower equilibrium constant of the Claus reaction. According to the concept of this reactor, reaction zones can be shifted from one place to another by varying the ratio between the driving forces for diffusion of the reactants. In this reactor setup, the mass transfer resistances in the gas phase were important, especially when a pressure difference over the membrane was vital but a moderate pressure difference did not affect the basic principle of this reactor.

In another work, a proposed mathematical model, based on the dusty gas model (DGM) and using the Claus reaction, described mass transport due to molecular diffusion and viscous flow combined with an instantaneous reversible reaction in a membrane reactor with opposite feed of reactants [40]. A comparative study concluded that the previously presented simplified model qualitatively predicted correct molar fluxes including the slip of reactants to the opposite sides of the membrane both in absence and presence of a moderate pressure difference over the membrane. The mathematically predicted conversion of sulfur at temperatures of 220°C and 268°C was presented to be 10% to 20% lower than the molar fluxes predicted by theoretical model. Conversions based on pressure difference demonstrated that the transport mechanism followed surface diffusion. No slip of reactant was observed until the pressure difference was up to 0.7 bar. A stainless steel membrane (pore diameter > 1 μm) was designed instead of using ceramic membrane due to the difficulty of impregnation of the catalyst, which would not be reproducible and controllable by means of homogeneity and activity for the iron-catalyzed oxidation of H_2S with separate feed of reactants [13]. In order to study the influence of diffusion and convection on the overall mass transfer, DGM, which predicts the conversion in the presence of a pressure difference, was implemented. This model predicted a substantial increase in conversion with the pressure difference. The conclusion was that a sintered metal membrane reactor acts equally as a ceramic membrane reactor does with separate feed of reactants.

5.2.5 Catalytic Combustion of Propane

For catalytic combustion of propane, a tubular support membrane made of α-alumina (nominal pore diameter: 0.7 μm; mean pore radius: 0.31 μm; porosity: 34.6%) was prepared; it was catalytically activated by applying a layer of Pt/γ-Al_2O_3 catalyst [41]. The behavior of the reactor in the absence of transmembrane pressure gradients was also studied. The reaction rate inside the membrane was compared to the transport rate of the reaction taking place in a small zone inside the membrane; no slip of propane and oxygen to their respective opposite sides occurred, which was attributed to the transport-controlled regime. A higher conversion was observed when propane was fed at the shell side with low concentrations and any slip of propane was prevented for concentrations lower than 35%. The stoichiometric excess shifted the reaction region toward the more active tube side of the membrane, thus enabling higher conversions than with a tube-side feed of propane.

In 1995, the same group [42] studied behavior of the reactor in the transport-controlled regime when pressure differences were applied over the membrane. A tubular catalytic membrane reactor consisting of a porous α-alumina with a γ-Al_2O_3 separation layer was fabricated and deposited on the pore walls of the membrane via the "urea method" [43]. Further, 1 wt% of platinum was deposited by the vacuum impregnation method. A propane–nitrogen mixture and air-containing stream were forced to permeate through a Pt/γ-Al_2O_3 catalytic membrane operating in a transport-controlled regime to provide close contact between the gas molecules and combustion sites, thus reducing the slip of reactants and providing high kinetics. The maximum conversion was achieved when propane was fed in stoichiometric excess (i.e., 7%); the reaction zone shifted toward the air side. Conversely, when propane was fed at a lower concentration (4%), the reaction zone tended to stand closer to the feed side, providing minimum conversion.

While the previous investigation [42] was limited to high-temperature operating conditions, the next study verified the transport of reactants and the conversion, as well as the influence of catalyst loading over the membrane on the reactor performance [44]. It was concluded that maximum conversion was achievable in the kinetics controlled regime when more catalyst was deposited in the membrane pores. Maximum propane conversions with minor slips of reactants through the membrane was possible when propane feed concentration and the transmembrane pressure gradient were not too high.

The concept of an opposite flow catalytic membrane reactor operating in a transport-controlled regime on the basis of nonisothermal modeling for catalytic combustion of propane has been reported in the literature [45]. A tubular catalytic membrane reactor (length: 100 mm) made of γ-Al$_2$O$_3$ with platinum (Pt) was enclosed in a cylindrical stainless steel module and separated by two chambers (i.e., shell side and tube side). The platinum content of the catalyst was maintained as 1 wt% to the weight of γ-Al$_2$O$_3$, whereas the γ-Al$_2$O$_3$ content of the membrane was kept at 4 wt% on the basis of the entire membrane weight. The influence of reaction kinetics on the reactor performance was observed only in the low-temperature and kinetics-controlled regime. The monodispersed model resulted in an unsatisfactory model performance, owing to position of the reaction zone due to uniform catalyst distribution. The bidispersed model was selected for the uneven catalyst distribution over the membrane. The isothermal model gave exact assessments of the achieved conversion only when the reactor was operating in the transport-controlled regime, but failed to predict the steady-state multiplicity that occurs at low temperatures.

5.2.6 Oxydehydrogenation of Propane to Propylene

V$_2$O$_5$/γ-Al$_2$O$_3$ catalyst-based conventional packed-bed reactor is used for oxydehydrogenation (ODH) of propane and is compared with other reactors such as a monolith-like reactor and catalytic membrane reactor [46]. A catalyst is prepared by deposition of vanadium pentoxide (V$_2$O$_5$) on the support of γ-Al$_2$O$_3$ powder in the form of a thin layer (0.2 wt%, thickness of the skin: 2 μm) and spread over on the internal side of a tubular porous ceramic membrane (height: 150 mm). The reactants are fed in cocurrent mode at the core side of the monolith-like reactor. In a packed-bed reactor, maximum conversion of propane is attained at the temperature of 450°C using 9.2 mol% isobutane, 1.6 mol% oxygen, and balanced helium as feedstock with propylene and carbon dioxide as products. The selectivity to propylene is obtained more than 60% at low residence time. It is reported that the selectivity is decreased with increasing propane conversion, whereas the conversion of propane and oxygen is increased when the residence time increases. The maximum selectivity of propylene, high turnover number, and maximum conversion of propane are achieved for monolith-like structures compared to packed-bed reactors. But, the catalytic membrane reactor shows an improvement in selectivity compared to both packed-bed reactors and monolith-like structures. But no improvement in selectivity is obtained with respect to the monolith-like configuration when a reverse CMR configuration is introduced because of quicker oxygen diffusion through the membrane from the shell to the core side.

The effects of different types of catalytic membranes and their feed arrangement are verified on the yield (propene) [47]. The influence of the back permeation of the reactant hydrocarbon and its prevention is mainly the focus. The conversion and selectivity of reactants are also measured where there is no transmembrane pressure gradient and no convective flow. Three types of tubular membranes of 200 nm in separation layer pore diameters with segregated feed mode are introduced. The membranes are (a) thin V/γ-Al$_2$O$_3$ layer on the inside of an α-Al$_2$O$_3$ support, (b) thin V/γ-Al$_2$O$_3$ layer on top of a SiO$_2$/α-Al$_2$O$_3$ support for better resistance to permeation, and (c) thin V/γ-Al$_2$O$_3$ layer on the tube side of a zeolite/α-Al$_2$O$_3$ support. Maximum selectivity is observed for the configuration in which oxygen (and helium) feed from the shell side due to lower partial pressure of oxygen in the active layer. It is noted that the segregated feed mode membrane offers high selectivity with respect to premixed feed membrane even if the diffusion resistance is low in the support layer at 550°C. However, high conversion values with decrease in selectivity have been observed for membranes having no zeolite crystals or silica within the support layer. The membrane made of silica with a higher diffusion resistance in the support layer provides higher conversion at both temperatures. Selectivity increases with increasing temperature, owing to strong reduction in back-diffusion and the distribution of oxygen within the support layer with significant diffusion resistance.

5.2.7 Other Catalytic Reactions

Two Pt/γ-Al$_2$O$_3$-activated porous membrane reactors (M1 and M2) with separate feed of reactants for the combustion of methane are prepared by the incipient wetness impregnation method by depositing platinum (1 wt% and 7 wt%) over the tubular α-alumina porous membrane [14]. The conversion is measured by means of different operating parameters such as operating temperature, feed (methane) concentration, type and amount of catalyst deposited, pressure difference applied over the membrane, time of operation, etc. In this study, the maximum specific heat power (15 kWm^{-2}) is obtained with negligible slip of methane to the air-feed side (an undesirable occurrence that reduces the advantages of this membrane reactor) and is close to M2, which is the most active and permeable membrane.

In another study, a Pd and Pt/SiO$_2$/α-Al$_2$O$_3$ membrane was prepared by depositing colloidal SiO$_2$ and by impregnating both Pd and Pt overnight, followed by calcination at 600°C for 2 h on α-Al$_2$O$_3$ membrane supports (inside diameter: 7 cm; outside diameter: 10 cm) for catalytic combustion of methane [48]. Methane and oxygen are fed separately to the tube and shell sides. Platinum membranes revealed lower CO production at above 580°C, but H$_2$ has never been detected in the outlet gases; no CO and H$_2$ are observed for palladium membranes.

The behavior of methanol conversion, influence of the membrane characteristics, sweep gas flow rate, effect of operating temperature, and different flux configurations for hydrogen production using the methanol steam reforming (MSR) reaction in a membrane reactor was verified by three different membranes (MR1, MR2, and MR3) [49]. Both cocurrent and countercurrent mode were applied. A comparative study between the fabricated membrane with a traditional membrane reactor (length: 250 mm; inside diameter: 6.7 mm) consisting of stainless steel tube has also been reported. MR1 is a Pd–Ag/TiO$_2$–Al$_2$O$_3$ asymmetric porous commercial membrane, whereas MR2 is an asymmetric porous membrane with a Pd–Ag deposit. MR3 is a dense, thin wall membrane tube with a Pd–Ag deposit. It was concluded that high methanol conversion, high hydrogen production, and low carbon monoxide selectivity are provided by MR3 in opposite-flow mode at high temperature and high sweep gas flow rate; which is superior to the other two membranes.

5.3 Fabrication of CMRs

In recent years, interest in the development of ceramic membranes for different catalytic reactions has increased due to their thermal and mechanical resistivity. Several types of support membranes, such as the disk shape, hollow fiber, and tubular, are well accepted in research fields. Nowadays, tubular membranes are favored by researchers because of their ability to handle process streams with high solids and high viscosity properties. The membrane area of tubular membrane per unit volume is small and fouling can easily be reduced by mechanical cleaning.

Usually ceramic membrane supports are formed by shaping a mixture of inorganic compounds with organic additives or inorganic binders followed by sintering. Several methods are used to prepare porous membranes from inorganic materials, such as phase separation, leaching, anodic oxidation, pyrolysis, particle dispersion, slip casting, and the sol-gel process [50]. A Pd-based tubular porous inorganic catalytic membrane reactor has been prepared and applied for water remediation by catalytic hydrogenation and H$_2$O$_2$ synthesis [51]. An MFI (silicalite-1 and ZSM-5)-based membrane presenting a nanocomposite structure within a 15 cm long porous alumina tube is prepared by hydrothermal synthesis of silica (Aerosil 380) and tetra-propylamonium hydroxide [52]. The incorporation of catalysts into membranes offers the opportunity not only to perform selective catalytic synthesis but also to separate reactants and/or products simultaneously. This review identifies the effect of use of different catalysts on different chemical process applications, given in Table 5.1.

During recent years, there have been a number of investigations dealing with the coating techniques for the deposition of active catalysts on the surface of the inorganic support membrane for catalytic membrane reactors such as physical vapor deposition (PVD) [63–66], chemical vapor deposition (CVD) [67], modified chemical vapor deposition [68], sol-gel coating [69–71], electroless plating and electroplating [72–74], the co-condensation and magnetic sputtering techniques [75], spray pyrolysis [76], etc.

CVD is a key technology in the field of catalytic membrane reactors to deposit active catalysts over the support membrane. The CVD process can be applied to modify membrane pore size, as this process generally provides good control of pore size. Though the process is quite time consuming and undesired compounds and impurities may be formed, causing reduction in the performance of the membrane [77], this process is widely used in the fabrication of products such as semiconductors, optoelectronics, optics, refractory fibers, filters, etc. [78].

In addition, the advantages of using CVD for membrane preparation are the simple, relatively inexpensive equipment and precursor materials for metal deposition [79,80]. In spite of using CVD for hydrocarbons, other membranes re-formed by CVD techniques, including palladium [81], SiC [82], and silica [83], are mostly used for separating gaseous mixtures. Theoretical analyses of CVD application to modify pore size were conducted earlier by using the population balance theory [84]. Another researcher has done the same study using percolation theory [85] and numerical analysis [86]. These studies have

TABLE 5.1

Use of Different Catalysts for Catalytic Membrane Reactors

Catalyst	Application	Ref.
Pd–Ru metal alloy	Selective and continuous methods of hydrogenation	[53]
Pd alloy	Cyclopentadiene to cyclopentene naphthalene to tetralin furan to tetrahydrofuran furfural to furfuryl alcohol	[54]
Pd/Ru/W	Dehydrogenation of 2-methyl-1-butene to isoprene	[55]
Pt/Si, Pt/Mo	Adsorption of ethene in the absence and presence of CO	[56]
Ga_2O_3/MoO_3	Oxidative dehydrogenation of propane	[57]
NiO/MgO	Catalytic partial oxidation of coke oven gas to syngas	[58]
Ni/Pd/Ag	H_2 production by low-pressure methane steam reforming	[59]
Ni-Nb-O	Oxidative dehydrogenation of ethane to ethylene	[60]
$CuO/ZnO/Al_2O_3$	Steam reforming of methanol	[61]
$CoMo/Al_2O_3$	Reaction of thiophene hydrodesulfurization	[62]

suggested that the pore size distribution of the membrane, the kinetics of the CVD reactions, and the transient kinetics of the modification process play important roles in the modification of pore size. A new technique, chemical-vapor infiltration (CVI), has been used by several groups. They prepared SiC membranes by CVI using SiH_2Cl_2 and C_2H_2, diluted with H_2; γ-Al_2O_3 tubes as supports [87]; and a SiC membrane using an α-alumina tube as the support. They used CVD of $(C_3H_7)_3SiH$ at 700°C–800°C [82].

The modified CVD co-deposition is another newly developed technique having several advantages compared to alternative techniques [88–90]. This process is also quite interesting in the application of platinum aluminide coatings for gas turbine hardware by CVD, co-deposition of aluminum, hafnium, zirconium, and silicon [91].

The electroless plating looks to be an attractive method as it requires very simple processing equipment and is able to coat a complexly shaped component with a layer of uniform thickness. Like CVD, it is a very prolonged process, but controlling thickness of a film is not easy by electroless plating. However, a significant advantage of this plating is that it is well matched to applications on available commercial membranes [92]. The catalytic dehydrogenation of isopropanol using a Cu/SiO_2 catalyst is performed in a Pd–Cu alloy membrane reactor. The membrane is prepared by electroless plating of alumina supports, followed by heat treatment after electroplating [93]. Two types of Ni–P alloy/ceramic membranes are prepared by the conventional electroless Ni-plating technique and the recrystallization technique for dehydrogenation of ethanol to acetaldehyde. Ethanol conversion is considerably higher in the former than in the membrane reactor using the recrystallized membrane [94].

In a recent study, a new, amorphous Ni–B alloy membrane created by electroless plating was employed for ethanol dehydrogenation. A Pd/Ag dense membrane on a mesoporous γ-Al_2O_3 membrane by electroless plating has been utilized for the catalytic dehydrogenation of methanol [95]. A combination of electroplating with the electroless chemical reduction using the patterned mask is a newly developed method. The advantages of this fabrication method are that the initial compositing between the polymer and platinum particles can be assured by the chemical reduction method and that the thickness of each electrode can be controlled easily and rapidly by electroplating [96]. Controlled thickness and porosity of a thin film can be accurately achieved by the sol-gel process. Composite membranes resulting from this process are generally microporous and mesoporous, on which permeation of gases is mainly controlled by surface transport and/or the Knudsen diffusion mechanism [82]. The preparation of microporous membranes by sol-gel modification of mesoporous γ-Al_2O_3 membranes was discussed in 1995 [97]. This approach has been successfully employed to vary the position of a narrow step distribution of Pt across the membrane radius [98]. The possibility of placing consecutive ceramic layers of different materials via sol-gel has also been discussed; each layer could be loaded with a different catalyst [99]. A V–P–O catalytic membrane prepared for the partial oxidation of butane via the sol-gel process shows the flexibility of this method to allow the synthesis of thin porous catalytic films that are difficult to achieve by other processes [100].

PVD is a less strenuous technique, offers faster deposition rates, and controls the film thickness but not the uniformity. To maintain the uniformity, rotation may be necessary; this depends on the shape of the component to be coated. In this process, the solid material to be deposited is first evaporated in a vacuum system, followed by condensation and deposition as a thin film on a cooler substrate. The magnetron sputtering technique contains a vacuum chamber comprising a target (a plate of the material to be deposited) and a substrate (the material). A high-energy ion strikes the conductive target, imparting momentum to atoms of the target and deposited on the target. Argon is generally used as sputtering gas. A combination of PVD and the sputtering technique for the deposition of palladium on the outer surface of commercial ceramic tubular membranes was investigated to conduct water gas shift reaction, but failed. A new economical and less time-consuming process (i.e., co-condensation) was then proposed and it looks to be more interesting for the researchers' purpose. By using this process, an apparently uniform Pd thin film (thickness 0.1 µm) that gives higher hydrogen flux through the layer has been obtained [77].

Another well-known method is spray pyrolysis, which is much simpler than other methods; a solution of metal salts is sprayed into a heated gas stream and pyrolyzed. This method has been applied for the formation of composite Pd membranes for hydrogen separation. The outer surface of the porous alumina alloy fiber is coated using $Pd(NO_3)_2$ and a $AgNO_3$ solution on a $H_2–O_2$ flame, to obtain a Pd–Ag alloy film [87]. In a recent work, synthesis of the $\gamma-Ga_2O_3–Al_2O_3$ solid solutions by spray pyrolysis has been examined to obtain spherical particles using an aqueous solution of $Ga(NO_3)_3$ and $Al(NO_3)_3$ in the presence of HNO_3. It is reported that spray pyrolysis gives amorphous products unless a sufficient thermal energy is supplied during the technique. That physical properties of the solid solutions are affected by the spray pyrolysis conditions has also been confirmed [101]. In another recent work, a new technique was introduced—ultrasonic spray analysis, which is a combination of spray analysis and aerosol-assisted CVD (AACVD), to prepare nitrogen-doped carbon nanotubes from mixtures of acetonitrile and imidazole. This has proved to be quite interesting [102].

5.4 Catalytic Membrane Reactors versus Traditional Reactors

Nowadays scientists and engineers are more attracted to CMRs compared to conventional reactors (plug flow reactor, continuous stirred tank reactor, etc.)

to remove/recover products involving both catalytic reaction and separation. There are a number of technical issues that make CMRs more interesting:

- Membrane reactors combine both reaction and separation simultaneously to increase conversion. One of the products of a given reaction is removed from the reactor through the membrane, forcing the equilibrium of the reaction to the right (according to Le Chatelier's principle), so that more product is produced.
- Conversion is enhanced, making the process more economical.
- Reaction rates increase.
- By-product formation is reduced.
- It is possible to separate molecules in a customized but cheap manner.
- Thermal damage is minimized.
- Recycling and lower exhausts are possible.
- Energy consumption is moderate.

Though the use of CMRs is advantageous and is involved in different applications, issues regarding manufacturing cost need to be addressed.

5.5 Possible Scopes of Further Research

Though the concept of the CMR is well established and involved in different applications, numerous issues can be addressed in further development of CMRs:

a. Selection of suitable materials for membrane fabrication
b. Parametric optimization (sintering temperature, applied pressure) of membrane preparation methods using a trial and error approach in experimentation
c. Evaluation of the best membrane module (tubular, hollow fiber, flat) and mode of application (dead end, cross flow, etc.)
d. Selection of suitable catalysts for different purposes
e. Evaluation of optimal operating conditions (preparation technique, calcination temperature) that enable offering higher yield
f. Choosing of simple coating techniques instead of using conventional processes like PVD, CVD, vacuum coating, etc.
g. Evaluation of the best membrane reactor configurations and modes of applications (extractor, distributor, and contactor)

A critical review from significant research findings conveys the following scopes for improvement in the fabrication of catalytic membrane reactors. Firstly, use of low-cost ceramic support rather than using alumina-based costly support should be stimulated to reduce the manufacturing cost of the catalytic membrane reactor for various industrial applications. Secondly, membrane fabrication, including catalyst preparation and coating techniques, needs to be simple and inexpensive with the least possible use of costly instruments. For this reason, conventional coating techniques are not a better choice as these they are very costly and not popular in industries. Hence, research in this direction must not be encouraged and a new manufacturing and coating technique must be discovered for the fabrication of catalytic membranes.

5.6 Membrane Contactors

Membrane contactors are nothing but a barrier between two phases involved in separations. In another sense, the term "membrane contactor" is used to classify membrane systems that are engaged to separate a desired species by keeping in contact with two phases. In the membrane contactor, the species are transported from one phase to another phase by diffusion, not by mixing or dispersion. The membranes used as membrane contactor can be microporous, symmetric, asymmetric, hydrophilic–hydrophobic, or composite (i.e., dense–microporous or hydrophilic–hydrophobic.

In the case of hydrophilic materials, the aqueous phase wets the membrane pores while the nonpolar/gas phase is restricted at the pore mouth (Figure 5.4a), whereas in the case of hydrophobic materials, the membrane

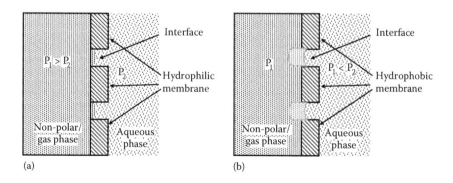

FIGURE 5.4
Interface/boundary between an aqueous phase and a nonpolar/gas phase in a (a) hydrophilic and (b) hydrophobic membrane.

can either be wetted by the nonpolar/gas phase while the aqueous phase cannot penetrate into the pores (Figure 5.4b). In this approach, operating pressure plays an important role, which has to be controlled carefully in order to avoid the mixing of two phases. In the case of hydrophilic membranes, the interface is observed at the pore mouth at the gas phase side and dispersion as drops between the phases is avoided by the pressure of the gas phase equal to or higher than the wetting phase pressure. In the case of hydrophobic membranes, the operating pressure of the aqueous phase has to be equal to or higher than the pressure of the wetting phase, which eliminates possibility of dispersion as drops of one phase into another phase. If the movement of the aqueous phase is restricted to get into the membrane pores, then the interfacial area can be recognized at the pore mouth. It is important to maintain the hydrophobicity of the material by controlling the critical value of pressure, called breakthrough pressure. If the breakthrough pressure is exceeded, the membrane loses its hydrophobicity and the aqueous phase starts to wet it [103–108]. Usually, breakthrough pressure depends on the pore radius, interface tension, and contact angle between the membrane and the fluid. It can be determined by using Laplace's equation, which offers a relationship between the largest pore radius of the membrane, $r_{p,max}$, and the breakthrough pressure ΔP_{start}:

$$\Delta P_{start} = \frac{2 \Theta \gamma \cos \theta}{r_{p,max}} \tag{5.1}$$

where
  Θ is a geometric factor associated with tortuosity factor
  γ is the interfacial tension
  θ is the solid–liquid contact angle

In all these cases, for simplicity, the tortuosity factor is considered as 1 because pores are considered to be straight and cylindrical.

The hydrophobicity of the membrane can also be influenced by the membrane structure and morphology. This can be controlled by using a composite membrane, in which a dense thin layer is coated over the microporous supported surface that restricts the penetration of the aqueous phase (Figure 5.5a). For asymmetric membranes, it is possible to keep the two phases in nondispersive contact at pressures higher than the breakthrough pressures. In these kinds of membranes, breakthrough pressure is inversely proportional to the pore sizes, as asymmetric membranes contain both large and small pores. In fact, there is a partial wetting of the membrane for the bigger pores, whereas the smaller pores continue to be aqueous free. This phenomenon indicates that the interfacial area can be established within the pores (Figure 5.5b). Figure 5.6 shows a contact between two liquid phases by means of a composite hydrophilic–hydrophobic membrane where the

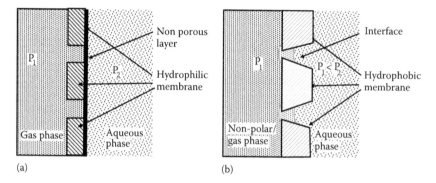

FIGURE 5.5
(a) Composite membrane with a dense layer coated on a supported microporous surface;
(b) interface between a nonpolar/gas phase and a aqueous phase in an asymmetric membrane.

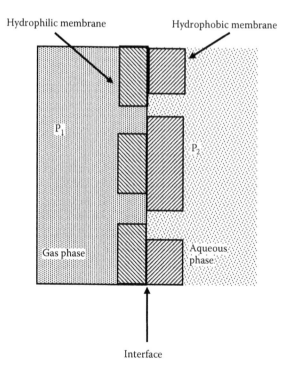

FIGURE 5.6
Interface between a nonpolar/gas phase and an aqueous phase in a composite hydrophilic–hydrophobic membrane.

aqueous phase wets the hydrophilic section and the nonaqueous phase enters the hydrophobic side.

The *advantages* of membrane contactors include:

- No dispersion between phases
- No loading, foaming, flooding
- Reduced size and weight
- No need of phase separation downstream
- High interfacial area in small volumes
- Wide range of operating flow rates
- Reaction and separation carried out simultaneously
- Flexible; easy scale-up

The *disadvantages* of membrane contactors include:

- Limited lifetime of the membrane
- Membrane fouling
- Channeling of the fluid
- Pretreatment before the process

5.6.1 Membrane Distillation

Membrane distillation (MD) is the only example of membrane contactor where the driving force is related to a temperature gradient across the membrane. It is a promising technology for different fields of industrial interest, such as wastewater treatment, desalination, ultrapure water production, concentration of nonvolatile components from aqueous solutions, etc. [109–111]. The advantages of MD compared to other popular and traditional separation processes are (1) 100% (theoretical) rejection of ions, macromolecules, colloids, cells, and other nonvolatiles; (2) lower operating temperatures than conventional distillation; (3) lower operating pressures than conventional pressure-driven membrane separation processes; (4) reduced chemical interaction between membrane and process solutions; (5) less demanding membrane mechanical property requirements; and (6) reduced vapor spaces compared to orthodox distillation processes. The main limitation of MD is the process solutions that must be aqueous and adequately dilute to prevent wetting of the hydrophobic microporous membrane. In addition to this, the volume of fluxes is also lower in MD than in other processes.

Figure 5.7 illustrates the basic four configurations:

   a. Direct contact membrane distillation (DCMD): An aqueous solution with a lower temperature is in direct contact with the permeate

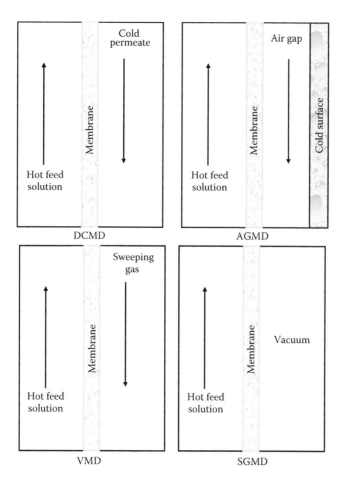

FIGURE 5.7
Four basic configuration of membrane distillation.

side of the membrane. The temperature difference across the membrane influences the vapor pressure difference. As a result, volatile molecules evaporate at the hot-liquid/vapor interface, transport across the membrane pores in vapor phase, and condense in the cold-liquid/vapor interface at the permeate side. DCMD has been extensively studied for desalination and concentration of aqueous solutions [112–118]. DCMD has the highest conductive heat loss among the four basic configurations, as membrane is the only barrier to separate the hot feed and cold permeate solutions.

b. Air gap membrane distillation (AGMD): A thin air gap is considered between the membrane and a condensation surface (normally a thin, dense polymer or metal film). The evaporated volatile molecules pass through both the membranes and the air gap and then

condense on the cold surface [119,120]. As the air gap offers a significant vapor transport resistance, the flux of a typical AGMD is lower
than DCMD or VMD configurations. Research has been carried out
on the multieffect or multistage membrane modules with improved
thermal efficiency [121,122].

c. Sweep gas membrane distillation (SGMD): A cold inert or sweep gas
sweeps through the permeate channel and collects vapor molecules
from the membrane surface. In most cases, the vapors are condensed
at the exterior surface of the membrane module by an external condenser [123], which enhances equipment cost.

d. Vacuum membrane distillation (VMD): Vacuum is applied at the
permeate side of the membrane module. To provide the driving
force, the applied vacuum must be lower than the saturation pressure of volatile molecules in the feed solution. Condensation may or
may not occur outside the membrane module [124–126]. SGMD and
VMD are often used to remove volatile organic compounds from
aqueous solutions [123].

5.6.2 Membrane Crystallizers

Crystallization is an outstanding procedure for purification of chemical species by solidification from a liquid mixture. The attractiveness of this method
is that the crystallization may operate at lower temperatures and requires
lower energy than other separation processes. Membrane crystallizers signify a specific application of membrane and osmotic distillation that per-
forms the same as membrane distillation but creates the partial pressure
gradient by sending the strip side an aqueous solution containing nonvolatile compounds at ambient temperature. In other words, a membrane crystallizer is an alternative technology of conventional methods like evaporation
for producing crystals from supersaturated solutions. Membrane crystallizers work on the same principles that control osmotic distillation operation.
The purpose of this concept is to achieve the crystallization of solutes by
eliminating water from the almost saturated feeds. This method avoids formation and precipitation of crystals on the membrane surface, which could
cause blockage of pores [127].

5.6.3 Membrane Emulsifier

Membrane emulsifiers engage both hydrophilic and hydrophobic membranes for creating microemulsions. Membrane emulsification has attracted
attention over the last 20 years in the field of membrane contactors, with possible applications in numerous fields. In the membrane emulsification process, a liquid phase (dispersed phase) is forced by applying pressure through
the membrane pores to form droplets at the permeate side of a membrane.

The droplets are then carried away by a continuous phase flowing across the membrane surface (Figure 5.8). The driving force is then the difference in pressure between the two phases. Theoretically, the membrane emulsification technique emphasizes two fundamental features: permeation of the dispersed phase through the pores of a membrane, and the dynamic mechanism of droplet detachment that rises from operating parameters such as cross-flow velocity, transmembrane pressure, morphological properties of the membrane (porosity, pore size, active pores), physical properties of the membrane (thickness, hydrophobic, or hydrophilic) and physicochemical properties of an emulsified system (density, surface tension, and viscosity).

The selection of membrane in this process strongly depends on the fact that the membrane surface should not wetted by the dispersed phase. For instance, for water/oil emulsions the membrane used is hydrophobic, whereas for oil/water emulsions the membrane used is hydrophilic. Under specific conditions, monodispersed emulsions can be produced using this technique [128]. In membrane emulsifiers, the flux mainly depends on the membrane resistance offered by the continuous phase and is directly proportional to the difference in pressure between the two phases. Small-scale applications such as drug delivery systems, food emulsions, and the production of monodispersed microspheres are the applications of membrane emulsifiers. Membrane emulsifiers offer advantages such as control of average droplet diameter by average membrane pore size and lower energy input. In addition—specifically in formulation of foods for production of oil/water emulsions (i.e., dressings, artificial milks, cream liqueurs), as well as for preparation of some water/oil emulsions (i.e., margarines and low-fat spreads)—the advantage of using membrane emulsifier is the minimum

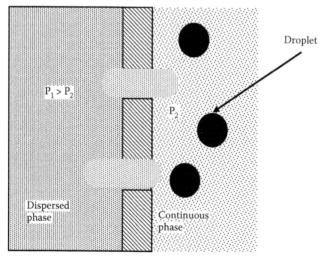

FIGURE 5.8
Schematic diagram of the membrane emulsification process.

shear forces on the physicochemical and molecular properties of the proteins. The main disadvantage of the membrane emulsification process is a low dispersed phase flux through the membrane.

5.6.4 Membrane Extractors

Membrane extractors comprise another significant segment of membrane contactors that can be used for carrying out liquid–liquid extractions normally conducted in mixer-settler, columns, and centrifugal devices. In this process, the driving force is the difference in concentration. The selection of membrane depends on the affinity of the species to be transported with the steams involved and minimum possibility of resistance offered by the membrane. Membrane extraction has progressed to be a feasible alternative to conventional sample preparation, and it has been used in the extraction of a wide range of compounds, including pesticides, pharmaceuticals, disinfection by-products, and metals [129]. It eases extraction without the mixing of two phases, thus rejecting emulsion formation. An important advantage of this process is that a sample and an extractant can be contacted continuously, thus providing the basis for a continuous, real-time process leading to automation and online connection to instruments. Consequently, membrane extraction has been directly interfaced with gas chromatography, liquid chromatography, mass spectrometry, ion chromatography, atomic spectroscopy, and capillary electrophoresis [130].

Conventional membrane extraction modules can be fabricated using hollow fibers having a tubular geometry or flat sheets. Usually hollow fiber modules are fabricated in a shell and tube design with multiple parallel fibers to offer high packing density. These modules provide higher surface area per unit volume compared to flat sheets [130].

Electromembrane extraction (EME) is a miniaturized liquid–liquid extraction technique developed for preparation of aqueous samples prior to analysis by chromatography, electrophoresis, mass spectrometry, and related techniques in analytical chemistry. EME involves the use of a small supported liquid membrane (SLM) persistent in the wall of a porous hollow fiber, and application of an electrical field across the SLM. EME provides high preconcentration and efficient sample cleanup. Furthermore, because the extraction is performed under the influence of an electrical field, the extraction selectivity can be governed by the direction and magnitude of the electrical field.

5.6.5 Phase Transfer Catalysis

Membrane contactors can also be used to carry out catalytic reactions. In this kind of technique, membranes are catalytically active and can be both hydrophobic and hydrophilic. The term *phase transfer catalysis* means that a compound of one phase can diffuse to the catalytic sites where the reaction is

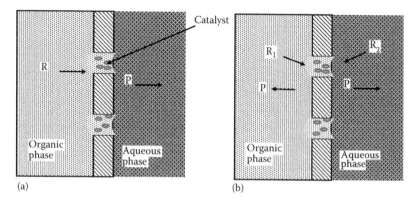

FIGURE 5.9
Schematic diagram of (a) phase transfer catalysis; (b) diffusion through separate feed of reactants and production.

taking place when both aqueous and organic phases are kept in contact, without mixing of the two streams (Figure 5.9a). The driving force for this system is a difference in concentration for both reactants and products. This concept is also applicable for the system containing reactants in both the streams. In this case, both the reactants will diffuse toward the catalytic sites and products can move toward both the streams, if the membrane is catalytically active itself. If either side of the membrane is coated with a catalytic layer, then both the reaction and separation can take place on the catalytic layer simultaneously. These types of membrane contactors can also be treated as catalytic membrane reactors as diffusion is taking place by separate feed of reactants and products (Figure 5.9b). Even three-phase reactions, where a gas and a liquid come in contact on the catalytic membrane, are also carried out using these types of membranes.

Phase transfer catalysis is a combination of both reaction and separation as in CMRs. The flux of reactants moving in the direction of the catalytic sites as well as the flux of products from the reaction region near the phases always depends on the membrane resistances and the resistances offered by the phases.

5.7 Summary

In this chapter, a detailed description of membrane reactors and membrane contactors has been provided, with several examples. The difference between traditional and advanced membrane reactor systems was also discussed. A new and innovative direction toward possible future scope was mentioned. The entire topic is summarized as follows:

- Membrane reactors—basic concepts, classification based on configurations A, B, C, D, E, F, G, and H
  - CMRs, IMRs, and combination of CMRs and IMRs
- The CMR and its novel applications—basic concepts of CMR are provided with several examples
  - Synergistic effect of separation and reaction
  - Membrane functions in CMR and applications—extractor, distributor and active contactor
    - Opposite flow mode catalytic membrane reactors
    - Hydrogen sulfide (H_2S) laden gas treatment
    - Catalytic combustion of propane
    - Oxydehydrogenation of propane to propylene
- Fabrication of CMRs—conventional methods with examples
  - Physical vapor deposition
  - Chemical vapor deposition
  - Modified chemical vapor deposition
  - Sol-gel coating
  - Electroless plating and electroplating
  - Co-condensation technique
  - Magnetic sputtering technique
  - Spray pyrolysis
- Advantages of catalytic membrane reactors
  - Enhanced conversion, making the process more economical
  - Increased reaction rates
  - Reduced by-product formation
  - Possibility of separating molecules in a customized but inexpensive manner
  - Minimization of thermal damage
  - Possibility for recycling and lower exhausts
  - Moderate energy consumption
- Membrane contactors—a barrier between two phases involved in separations
- Advantages of membrane contactors
  - No dispersion between phases
  - No loading, foaming, flooding
  - Reduced size and weight
  - No need of phase separation downstream

- High interfacial area in small volumes
- Wide range of operating flow rates
- Reaction and separation carry out simultaneously
- Flexible; easy scale-up
- Disadvantages of membrane contactors
  - Limited lifetime of the membrane
  - Membrane fouling
  - Channeling of the fluid
    - Pretreatment before the process
- Different types of membrane contactors
  - Membrane distillation—driving force is the temperature gradient across the membrane.
  - Membrane crystallizers—crystallization may operate at lower temperatures and requires lower energy than other separation processes.
  - Membrane emulsifier—driving force is the difference of pressure between the two phases.
  - Membrane extractors—driving force is the difference in concentration.
  - Phase transfer catalysis—a compound of one phase can diffuse to the catalytic sites where the reaction is taking place when both aqueous and organic phases are kept in contact, without mixing of the two streams. The driving force for this system is a difference in concentration for both reactants and products.

References

1. E. Gobina, R. Hughes, Ethane dehydrogenation using a high-temperature catalytic membrane reactor. *Journal of Membrane Science*, 90 (1–2) (1994) 11–19.
2. S. Uemiya, N. Sato, H. Ando, E. Kikuchi, The water gas shift reaction assisted by a palladium membrane reactor. *Industrial Engineering & Chemistry Research*, 30 (3) (1991) 585–589.
3. Y.V. Gokhale, R.D. Noble, J.L. Falconer, Effects of reactant loss and membrane selectivity on a dehydrogenation reaction in a membrane-enclosed catalytic reactor. *Journal of Membrane Science*, 103 (3) (1995) 235–242.
4. F. Tiscarño-Lechuga, C.G. Hill Jr., M.A. Anderson, Effect of dilution in the experimental dehydrogenation of cyclohexane in hybrid membrane reactors. *Journal of Membrane Science*, 118 (1) (1996) 85–92.

5. N. Itoh, Simulation of a bifunctional palladium membrane reactor. *Journal of Chemical Engineering Japan*, 23 (1) (1990) 81–87.
6. E. Gobina, R. Hughes, Reaction coupling in catalytic membrane reactors. *Chemical Engineering Science*, 51 (11) (1996) 3045–3050.
7. A. Santos, C. Finol, J. Coronas, M. Menendez, J. Santamaria, Reactor engineering studies of methane oxidative coupling on a Li/Mgo catalyst. *Studies in Surface Science and Catalysis*, 81 (C) (1994) 171–176.
8. D. Lafarga, J. Santamaria, M. Menendez, Methane oxidative coupling using porous ceramic membrane reactors—I. Reactor development. *Chemical Engineering Science*, 49 (12) (1994) 2005–2013.
9. R. Mallada, M. Menendez, J. Santamaria, *Proceedings of the Third International Conference on Catalysis in Membrane Reactors*, Copenhagen, September 8–10 (1998).
10. E. Xue, J. Ross, *Proceedings of the Third International Conference on Catalysis in Membrane Reactors*, Copenhagen, September 8–10 (1998).
11. H.J. Sloot, A non-permselective membrane reactor for catalytic gas phase reactions. PhD thesis, University of Twente, the Netherlands (1991).
12. J.W. Veldsink, G.F. Versteeg, W.P.M. van Swaaij, *Proceedings of the First International Workshop on Catalytic Membranes*, Lyon-Villeurbanne, France, September 26–28 (1994).
13. H.W.J.P. Neomagus, W.P.M. van Swaaij, G.F. Versteeg, The catalytic oxidation of H_2S in a stainless steel membrane reactor with separate feed of reactants. *Journal of Membrane Science*, 148 (2) (1998) 147–160.
14. H.W.J.P. Neomagus, G. Saracco, H.F.W. Wessel, G.F. Versteeg, The catalytic combustion of natural gas in a membrane reactor with separate feed of reactants. *Chemical Engineering Journal*, 77 (3) (2000) 165–177.
15. M.P. Harold, V.T. Zaspalis, K. Keizer, A.J. Burgraaf, Intermediate product yield enhancement with a catalytic inorganic membrane—I. Analytical model for the case of isothermal and differential operation. *Chemical Engineering Science*, 48 (15) (1993) 2705–2725.
16. J. Herguido, D. Lafarga, M. Menendez, J. Santamaria, C. Guimon, Characterization of porous ceramic membranes for their use in catalytic reactors for methane oxidative coupling. *Catalysis Today*, 25 (3–4) (1995) 263–269.
17. J. Coronas, J. Santamaria, Catalytic reactors based on porous ceramic membranes. *Catalysis Today*, 51 (1999) 377–389.
18. P. Cini, M.P. Harold, Experimental study of the tubular multiphase catalyst. *AIChE Journal*, 37 (7) (1991) 997–1008.
19. J. Peureux, M. Torres, H. Mozzanega, A. Giroir-Fendler, J.A. Dalmon, Nitrobenzene liquid-phase hydrogenation in a membrane reactor. *Catalysis Today*, 25 (3–4) (1995) 409–415.
20. G. Saracco, V. Specchia, Catalytic ceramic filters for flue gas cleaning. 2. Catalytic performance and modeling thereof. *Industrial Engineering & Chemistry Research*, 34 (4) (1995) 1480–1487.
21. G. Saracco, S. Specchia, V. Specchia, Catalytically modified fly-ash filters for NOx reduction with NH_3. *Chemical Engineering Science*, 51 (24) (1996) 5289–5297.
22. H.P. Hsieh, *Inorganic membranes for separation and reaction*, Elsevier Science B.V., Amsterdam, the Netherlands (1996).
23. A. Julbe, D. Farrusseng, C. Guizard, Porous ceramic membranes for catalytic reactors—Overview and new ideas. *Journal of Membrane Science*, 181 (2001) 3–20.

24. J.G. Sanchez Marcano, T.T. Tsotsis, *Catalytic membranes and membrane reactors*, Wiley-VCH Verlag GmbH, Weinheim, Germany (2002).
25. I. Pinnau, L.G. Toy, Solid polymer electrolyte composite membranes for olefin–paraffin separation. *Journal of Membrane Science*, 184 (2001) 39–48.
26. H. Weyten, K. Keizer, A. Kinoo, J. Luyten, R. Leysen, Dehydrogenation of propane using a packed-bed catalytic membrane reactor. *AIChE Journal*, 43 (7) (1997) 1819–1827.
27. H. Weyten, J. Luyten, K. Keizer, L. Willems, R. Leysen, Membrane performance: The key issues for dehydrogenation reactions in a catalytic membrane reactor. *Catalysis Today*, 56 (1–3) (2000) 3–11.
28. C.Y. Tsai, Y.H. Ma, W.R. Moser, A.G. Dixon, Modelling and simulation of a non-isothermal catalytic membrane reactor. *Chemical Engineering Communications*, 134 (1) (1995) 107–132.
29. P. Chanaud, A. Julbe, A. Larbot, C. Guizard, L. Cot, H. Borges, A.G. Fendler, C. Mirodatos, Catalytic membrane reactor for oxidative coupling of methane. Part 1: Preparation and characterisation of LaOCl membrane. *Catalysis Today*, 25 (3–4) (1995) 225–230.
30. M.P. Pina, M. Menendez, J. Santamaria, The Knudsen-diffusion catalytic membrane reactor: An efficient contactor for the combustion of volatile organic compounds. *Applied Catalysis B: Environment*, 11 (1) (1996) L19–L27.
31. C.R. Binkerd, Y.H. Ma, W.R. Moser, A.G. Dixon, An experimental study of the oxidative coupling of methane in porous ceramic radial-flow catalytic membrane reactors, *in Proceedings of the Fourth International Congress on Inorganic Membranes*, Gatlinburg, TN, July 14–18, 441 (1996).
32. C.K. Lambert, R.D. Gonzalez, Activity and selectivity of a Pd/γ-Al$_2$O$_3$ catalytic membrane in the partial hydrogenation reactions of acetylene and 1,3-butadiene. *Catalysis Letters*, 57 (1–2) (1991) 1–7.
33. W.A. Jacoby, P.C. Maness, D.M. Blake, E.J. Wolfrum, *Proceedings of the Third International Conference on Catalysis in Membrane Reactors*, Copenhagen, 8–10 September, Paper O13, (1998).
34. W.F. Maier, C. Lange, I. Tilgner, B. Tesche, Effects of membrane catalyst on the activity and selectivity of hydrogenation reactions, in *Proceedings of the Third International Conference on Catalysis in Membrane Reactors*, P.E. Hojlund, N. Haldor, A.S. Topsoe (eds.) p. 22, Copenhagen, Denmark, September (1998).
35. X. Dong, W. Jin, N. Xua, K. Li, Dense ceramic catalytic membranes and membrane reactors for energy and environmental applications. *Chemical Communications*, 47 (2011) 10886–10902.
36. V.T. Zaspalis, W. van Praag, K. Keizer, J.G. van Ommen, J.R.H. Ross, A.J. Burgraaf, Reactor studies using vanadia modified titania and alumina catalytically active membranes for the reduction of nitrogen oxide with ammonia. *Applied Catalysis*, 74 (1991) 249–260.
37. J.A. Dalmon, Catalytic membrane reactors, in *Handbook of heterogeneous catalysis*, G. Ertl, H. Knözinger, J. Weitkamp (eds.) (Chapter 9.3) Wiley-VCH Publishers, Weinheim, Germany (1997).
38. H.J. Sloot, G.F. Versteeg, W.P.M. van Swaaij, A non-permselective membrane reactor for chemical processes normally requiring strict stoichiometric feed of reactants. *Chemical Engineering Science*, 45 (1990) 2415–2421.
39. H.F. Mark, D.F. Othmer, C.G. Overberger, G.T. Seaborg, K. Othmer, *Encyclopedia of chemical technology*, vol. 22, John Wiley & Sons, New York (1983).

40. H.J. Sloot, C.A. Smolders, W.P.M. van Swaaij, G.F. Versteeg, High-temperature membrane reactor for catalytic gas–solid reactions. *AIChE Journal*, 38 (1992) 887–900.
41. G. Saracco, J.W. Veldsink, G.F. Versteeg, W.P.M. van Swaaij, Catalytic combustion of propane in a membrane reactor with separate feed of reactants—I. Operation in absence of trans-membrane pressure gradients. *Chemical Engineering Science*, 50 (12) (1995) 2005–2015.
42. G. Saracco, J.W. Veldsink, G.F. Versteeg, W.P.M. van Swaaij, Catalytic combustion of propane in a membrane reactor with separate feed of reactants—II. Operation in presence of trans-membrane pressure gradients. *Chemical Engineering Science*, 50 (17) (1995) 2833–2841.
43. L. Gordon, M.L. Salutsky, H.H. Willard, *Precipitation from homogeneous solutions*, Chapman & Hall, London (1959).
44. G. Saracco, J.W. Veldsink, G.F. Versteeg, W.P.M. van Swaaij, Catalytic combustion of propane in a membrane reactor with separate feed of reactants—III. Role of catalyst load on reactor performance. *Chemical Engineering Science*, 147 (1) (1996) 29–42.
45. G. Saracco, V. Specchia, Catalytic combustion of propane in a membrane reactor with separate feed of reactants—IV. Transition from the kinetics to the transport-controlled regime. *Chemical Engineering Science*, 55 (2000) 3979–3989.
46. G. Capannelli, E. Carosini, F. Cavani, O. Monticelli, F. Trifiro, Comparison of the catalytic performance of V_2O_5/γ-Al_2O_3 in the oxi-dehydrogenation of propane to propylene in different reactor configurations: i) packed-bed reactor, ii) monolith-like reactor and iii) catalytic membrane reactor. *Chemical Engineering Science*, 51 (10) (1996) 1817–1826.
47. M.J. Alfonso, A. Julbe, D. Farrusseng, M. Menéndez, J. Santamaría, Oxidative dehydrogenation of propane on V/Al_2O_3 catalytic membranes. Effect of the type of membrane and reactant feed configuration. *Chemical Engineering Science*, 54 (1999) 1265–1272.
48. M. Murru, A. Gavriilidis, Catalytic combustion of methane in non-permselective membrane reactors with separate reactant feeds. *Chemical Engineering Journal*, 100 (2004) 23–32.
49. A. Basile, S. Tosti, G. Capannelli, G. Vitulli, A. Iulianelli, F. Gallucci, E. Drioli, Co-current and counter-current modes for methanol steam reforming membrane reactor: Experimental study. *Catalysis Today*, 118 (2006) 237–245.
50. J. Luyten, A. Buekenhoudt, W. Adriansens, J. Cooymans, H. Weyten, F. Servaes, R. Leysen, Preparation of $LaSrCoFeO_{3-x}$ membranes. *Solid State Ionics*, 135 (1–4) (2000) 637–642.
51. G. Centri, R. Dittmeyer, S. Perathoner, M. Reif, Tubular inorganic catalytic membrane reactors: Advantages and performance in multiphase hydrogenation reactions. *Catalysis Today*, 79–80 (2003) 139–149.
52. J.-A. Dalmon, A. Cruz-Lo´pez, D. Farrusseng, N. Guilhaume, E. Iojoiu, J.-C. Jalibert, S. Miachon, C. Mirodatos, A. Pantazidis, M.R-. Dassonneville, Y. Schuurman, A.C. van Veen, Oxidation in catalytic membrane reactors. *Applied Catalysis A: General*, 325 (2007) 198–204.
53. M.M. Ermilova, L.S. Morozov, N.V. Smith, V.M. Gryaznov, Effect of catalytic hydrogenation on hydrogen transfer through the membrane of palladium–ruthenium alloy. *Nature Chemistry*, 68 (7C) (1994) 1211–1214.

54. V.M. Gryaznov, A.N. Karavanov, Hydrogenation and dehydrogenation of organic compounds on membrane catalysts (review). *Pharmaceutical Chemistry Journal*, 7 (1979) S74–S78.

55. V.S. Smirnov, V.M. Gryaznov, V.I. Lebedeva, A.P. Mishchenko, V.P. Polyakova, E.M. Savitskii, Catalysts for dehydrogenation, dehydrocyclization, and dehydroalkylation of hydrocarbons. German Öffentlichung, DE 2015248 A (1971) 19710218 (patent).

56. S.D. Jackson, B.M. Glanville, J. Willis, G.D. McLellan, G. Webb, R. B. Moyes, S. Simpson, P.B. Wells, R. Whyman, Supported metal catalysts: Preparation, characterisation, and function III. The adsorption of hydrocarbons on platinum catalysts. *Journal of Catalysis*, 139 (1993) 221–233.

57. Ž.S. Kotanjac, M. vanSint Annaland, J.A.M. Kuipers, A packed bed membrane reactor for the oxidative dehydrogenation of propane on a Ga_2O_3/MoO_3 based catalyst. *Chemical Engineering Science*, 65 (2010) 441–445.

58. Z. Yang, W. Ding, Y. Zhang, X. Lu, Y. Zhang, P. Shen, Catalytic partial oxidation of coke oven gas to syngas in an oxygen permeation membrane reactor combined with NiO/MgO catalyst. *International Journal of Hydrogen Energy*, 35 (2010) 6239–6247.

59. A. Iulianelli, G. Manzolini, M. De Falco, S. Campanari, T. Longo, S. Liguori, A. Basile, H2 production by low pressure methane steam reforming in a Pd/Ag membrane reactor over a Ni-based catalyst: Experimental and modelling. *International Journal of Hydrogen Energy*, 35 (2010) 11514–11524.

60. M.L. Rodriguez, D.E. Ardissone, E. Heracleous, A.A. Lemonidou, E. Lopez, M.N. Pedernera, D.O. Borio, Oxidative dehydrogenation of ethane to ethylene in a membrane reactor: A theoretical study. *Catalysis Today*, 157 (2010) 303–309.

61. S. Sa´, J.M. Sousa, A. Mendes, Steam reforming of methanol over a CuO/ZnO/Al_2O_3 catalyst part II: A carbon membrane reactor. *Chemical Engineering Science*, 66 (2011) 5523–5530.

62. Ch. Vladov, L. Petrov, B. Ytua, Structure and activity of a CoMo/Al_2O_3 catalyst upon modification by gamma irradiation. *Applied Catalysis A: General*, 94 (1993) 205–213.

63. V. Jayaraman, Y.S. Lin, M. Pakala, R.Y. Lin, Fabrication of ultrathin metallic membranes on ceramic supports by sputter deposition. *Journal of Membrane Science*, 99 (1995) 89–100.

64. E. Gobina, R. Hughes, *Proceedings of the First International Workshop on Catalytic Membranes*, Lyon-Villeurbanne, France, September 26–28, Cl7 (1994).

65. V.M. Linkov, R.D. Sanderson, E.P. Jacobs, S.P.J. Smith, *Proceedings of the First International Workshop on Catalytic Membranes*, Lyon-Villeurbanne, France, September 26–28, C24 (1994).

66. A.L. Athayde, R.W. Baker, P. Nguyen, Metal composite membranes for hydrogen separation. *Journal of Membrane Science*, 94 (1) (1994) 299–311.

67. S. Yan, H. Maeda, K. Kusakabe, S. Morooka, Thin palladium membrane formed in support pores by metal-organic chemical vapor deposition method and application to hydrogen separation. *Industrial Engineering & Chemistry Research*, 33 (3) (1994) 616–622.

68. C.E. Megris, J.H.E. Glezer, Synthesis of hydrogen-permselective membranes by modified chemical vapor deposition. Microstructure and permselectivity of silica/carbon/Vycor membranes. *Industrial Engineering & Chemistry Research*, 31 (5) (1992) 1293–1299.

69. V. Parvulescu, A. Julbe, L. Cot, L. Popescu, *Proceedings of the First International Workshop on Catalytic Membranes*, Lyon-Villeurbanne, France, September 26–28, P6 (1994).

70. K. Kusakabe, K. Ichiki, S. Morooka, Separation of CO_2 with $BaTiO_3$ membrane prepared by the sol-gel method. *Journal of Membrane Science*, 95 (2) (1994) 171–177.

71. A. Julbe, C. Guizard, A. Larbot, L. Cot, A. Giroir-Fendler, The sol-gel approach to prepare candidate microporous inorganic membranes for membrane reactors. *Journal of Membrane Science*, 77 (2–3) (1993) 137–153.

72. M.L. Chou, N. Manning, H. Chen, Deposition and characterization of thin electroless palladium films from newly developed baths. *Thin Solid Films*, 213 (1) (1992) 64–71.

73. J. Shu, B.P.A. Grandjean, E. Ghali, S. Kakiaguine, Simultaneous deposition of Pd and Ag on porous stainless steel by electroless plating. *Journal of Membrane Science*, 77 (2–3) (1993) 181–195.

74. E. Kikuchi, *Proceedings of the First International Workshop on Catalytic Membranes*, Lyon-Villeurbanne, France, September 26–28, Cl9 (1994).

75. G. Capannelli, A. Bottino, G. Gao, A. Grasso, A. Servida, G. Vitulli, A. Mastrantuono, R. Lazzaroni, P. Salvadori, Porous $Pt/\gamma\text{-}Al_2O_3$ catalytic membrane reactors prepared using mesitylene solvated Pt atoms. *Catalysis Letters*, 20 (1993) 287–297.

76. Z.Y. Li, H. Maeda, K. Kusakabe, S. Morooka, H. Anzai, S. Akiyama, Preparation of palladium–silver alloy membranes for hydrogen separation by the spray pyrolysis method. *Journal of Membrane Science*, 78(3) (1993) 247–254.

77. A. Basile, E. Drioli, F. Santella, V. Violante, G. Capannelli, G. Vitulli, A study on catalytic membrane reactors for water gas shift reaction. *Gas Separation & Purification*, 10 (1) (1996) 53–61.

78. Y.Y. Li, T. Nomura, A. Sakoda, M. Suzuki, Fabrication of carbon coated ceramic membranes by pyrolysis of methane using a modified chemical vapor deposition apparatus. *Journal of Membrane Science*, 197 (1–2) (2002) 23–35.

79. G. Xomeritakis, Y.S. Lin, Fabrication of a thin palladium membrane supported in a porous ceramic substrate by chemical vapor deposition. *Journal of Membrane Science*, 120 (2) (1996) 261–272.

80. G. Xomeritakis, Y.S. Lin, CVD synthesis and gas permeation properties of thin palladium/alumina membranes. *AIChE Journal*, 44 (1998) 174–183.

81. G. Xomeritakis, Y.S. Lin, Fabrication of thin metallic membranes by MOCVD and sputtering. *Journal of Membrane Science*, 133 (2) (1997) 217–230.

82. B.K. Sea, K. Ando, K. Kusakabe, S. Morooka, Separation of hydrogen from steam using a SiC-based membrane formed by chemical vapor deposition of tri-isopropylsilane. *Journal of Membrane Science*, 146 (1) (1998) 73–82.

83. K. Kuraoka, Z. Shugen, K. Okita, T. Kakitani, T. Yazawa, Permeation of methanol vapor through silica membranes prepared by the CVD method with the aid of evacuation. *Journal of Membrane Science*, 160 (1) (1999) 31–39.

84. Y.S. Lin, A theoretical analysis on pore size change of porous ceramic membranes after modification. *Journal of Membrane Science*, 79 (1) (1993) 55–64.

85. M. Tsapatsis, G. Gavalas, Modeling of SiO_2 deposition in porous Vycor: Effects of pore network connectivity. *AIChE Journal*, 43 (1997) 1849–1860.

86. L.L. Lee, L.C. Hong, L.S. Hong, D.S. Tsai, Pore structure modification by chemical vapor deposition in inorganic membrane-numerical analysis. *Journal of Chinese Institute of Chemical Engineering*, 30 (1999) 105–115.

87. Y. Takeda, N. Shibata, Y. Kubo, SiC coating on porous γ-Al$_2$O$_3$ using alternative-supply CVI method. *Journal of Ceramic Society Japan,* 109 (2001) 305–309.
88. J. Schaeffer, B. Gupta, NASA technical brief, published jointly by National Aeronautics and Space Administration (NASA) and Associated Business Publications Co, New York (1999).
89. L. Viterna, *Advanced materials and processes,* 20 pp., ASM International, Materials Park, OH (1999).
90. D.J. Rigney, R. Darolia, W.S. Walston, European patent application #EP-1,010,744-A1 10 July (1999).
91. B.M. Warnes, Reactive element modified chemical vapor deposition low activity platinum aluminide coatings. *Surface Coating Technology,* 146–147 (2001) 7–12.
92. R.E. Buxbaum, T.L. Marker, Hydrogen transport through non-porous membranes of palladium-coated niobium, tantalum and vanadium. *Journal of Membrane Science,* 85 (1) (1993) 29–38.
93. D.W. Mouton, J.N. Keuler, L.L. Lorenzen, *Proceedings of the Sixth International Congress on Inorganic Membranes,* Montpellier, France, June 26–30, 137 (2000).
94. B.S. Liu, W.L. Dai, G.H. Wu, J.F. Deng, Amorphous alloy/ceramic composite membrane: Preparation, characterization and reaction studies. *Catalysis Letters,* 49 (3–4) (1997) 181–188.
95. W. Lefu, J. Hongbing, Z. Lie, *Proceedings of the Fourth International Conference on Catalysis in Membrane Reactors,* 63 pp. (2000).
96. J.H. Jeon, S.W. Yeom, II-K. Oh, Fabrication and actuation of ionic polymer metal composites patterned by combining electroplating with electroless plating. *Composites Part A—Applied Science and Manufacturing,* 39 (2008) 588–596.
97. R.S.A. de Lange, J.H.A. Hekkink, K. Keizer, A.J. Burggraaf, Formation and characterization of supported microporous ceramic membranes prepared by sol-gel modification techniques. *Journal of Membrane Science,* 99 (1) (1995) 57–75.
98. K.L. Yeung, R. Aravind, R.J.X. Zawada, J. Szegner, G. Gao, A. Varma, Nonuniform catalyst distribution for inorganic membrane reactors: Theoretical considerations and preparation techniques. *Chemical Engineering Science,* 49 (24) part A (1994) 4823–4838.
99. A.S. Michaels, Seventh ESMST Summer School, University of Twente, the Netherlands (1989).
100. D. Farrusseng, A. Julbe, D. Cot, C. Guizard, S. Mota, J.C. Volta, *Proceedings of the Third International Conference on Catalysis in Membrane Reactors,* Copenhagen, September 8–10 (1998).
101. T. Watanabe, Y. Miki, T. Masuda, H. Deguchi, H. Kanai, S. Hosokawa, K. Wada, M. Inoue, Synthesis of γ-Ga$_2$O$_3$–Al$_2$O$_3$ solid solutions by spray pyrolysis method. *Ceramics International,* 37 (2011) 3183–3192.
102. J. Liu, Y. Zhang, M.I. Ionescu, R. Li, X. Sun, Nitrogen-doped carbon nanotubes with tunable structure and high yield produced by ultrasonic spray pyrolysis. *Applied Surface Science,* 257(17) (2011) 7837–7844.
103. E. Drioli, A. Criscuoli, Microporous inorganic and polymeric membranes as catalytic reactors and membrane contactors, in N. Kanellopoulus (ed.), *Recent advances in gas separation by microporous membranes,* Membrane science and technology series, 6, Elsevier, Amsterdam (2000) 497–510.

104. A. Criscuoli, E. Curcio, E. Drioli, Polymeric membrane contactors, in S.G. Pandalai (ed.), Recent research developments in applied polymer science, Transworld Research Network publication by Research Signspot, ISBN:81-7895-102-9, Kerala, 37/66 (2), 7 (2003) 1–21.

105. R. Prasad, K.K. Sirkar, Membrane based solvent extraction, in W.S.W Ho and K.K. Sirkar (eds.), *Membrane handbook*, Chapman and Hall, New York (1992) 727–763.

106. E. Drioli, A. Criscuoli, E. Curcio, Membrane contactors and catalytic membrane reactors in process intensification. *Chemical Engineering Technology*, 26 (9) (2003) 975–981.

107. H. Kreulen, C.A. Smolders, G.F. Versteeg, W.P.M. van Swaaij, Determination of mass transfer rates in wetted and non-wetted microporous membranes. *Chemical Engineering Science*, 48 (1993) 2093–2102.

108. A. Malek, K. Li, W.K. Teo, Modeling of microporous hollow fiber membrane modules operated under partially wetted conditions. *Industrial Engineering and Chemistry Research*, 36 (1996) 784–793.

109. A.M. Alklaibi, N. Lior, Membrane-distillation: Status and potential. *Desalination* 171 (2004) 111–131.

110. M.S. El-Bourawi, Z. Ding, R. Ma, M. Khayet, A framework for better understanding membrane distillation separation process. *Journal of Membrane Science*, 285 (2006) 4–29.

111. A. Alkhudhiri, N. Darwish, N. Hilal, Membrane distillation: A comprehensive review. *Desalination*, 287 (2012) 2–18.

112. K.W. Lawson, D.R. Lloyd, Membrane distillation. *Journal of Membrane Science*, 124 (1997) 1–25.

113. M. Khayet, T. Matsuura, Preparation and characterization of polyvinylidene fluoride membranes for membrane distillation. *Industrial and Engineering Chemistry Research*, 40 (2001) 5710–5718.

114. L. Song, Z. Ma, X. Liao, P.B. Kosaraju, J.R. Irish, K.K. Sirkar, Pilot plant studies of novel membranes and devices for direct contact membrane distillation-based desalination. *Journal of Membrane Science*, 323 (2008) 257–270.

115. M. Gryta, M. Tomaszewska, J. Grzechulska, A.W. Morawski, Membrane distillation of NaCl solution containing natural organic matter. *Journal of Membrane Science*, 181 (2001) 279–287.

116. K.Y. Wang, S.W. Foo, T.S. Chung, Mixed matrix PVDF hollow fiber membranes with nanoscale pores for desalination through direct contact membrane distillation. *Industrial & Engineering Chemistry Research*, 48 (2009) 4474–4483.

117. S. Bonyadi, T.S. Chung, Flux enhancement in membrane distillation by fabrication of dual layer hydrophilic–hydrophobic hollow fiber membranes. *Journal of Membrane Science*, 306 (2007) 134–146.

118. M. Su, M.M. Teoh, K.Y. Wang, J. Su, T.S. Chung, Effect of inner-layer thermal conductivity on flux enhancement of dual-layer hollow fiber membranes in direct contact membrane distillation. *Journal of Membrane Science*, 364 (2010) 278–289.

119. G. Lewandowicz, W. Białas, B. Marczewski, D. Szymanowska, Application of membrane distillation for ethanol recovery during fuel ethanol production. *Journal of Membrane Science*, 375 (2011) 212–219.

120. R. Thiruvenkatachari, M. Manickam, T.O. Kwon, I.S. Moon, J.W. Kim, Separation of water and nitric acid with porous hydrophobic membrane by airgap membrane distillation (AGMD). *Separation Science and Technology*, 41 (2006) 3187–3199.

121. E. Guillén-Burrieza, J. Blanco, G. Zaragoza, D.-C. Alarcón, P. Palenzuela, M. Ibarra, W. Gernjak, Experimental analysis of an airgap membrane distillation solar desalination pilot system. *Journal of Membrane Science,* 379 (2011) 386–396.

122. Y. Qin, Y. Wu, L. Liu, D. Cui, Y. Zhang, D. Liu, K. Yao, Y. Liu, A. Wang, W. Li, Multi-effect membrane distillation process for desalination and concentration of aqueous solutions of non-volatile or semi-volatile solutes. AIChE Annual Meeting, Salt Lake City, UT, 2010.

123. Z.L. Xie, T. Duong, M. Hoang, C. Nguyen, B. Bolto, Ammonia removal by sweep gas membrane distillation. *Water Research,* 43 (2009) 1693–1699.

124. S. Al-Asheh, F. Banat, M. Qtaishat, M. Al-Khateeb, Concentration of sucrose solutions via vacuum membrane distillation. *Desalination,* 195 (2006) 60–68.

125. J. Liu, C. Wu, X. Lü, Heat and mass transfer in vacuum membrane distillation. *CIESC,* 62 (2011) 908–915.

126. B. Li, K.K. Sirkar, Novel membrane and device for vacuum membrane distillation-based desalination process. *Journal of Membrane Science,* 257 (2005) 60–75.

127. E. Curcio, A. Criscuoli, E. Drioli, Membrane crystallizers. *Industrial Engineering & Chemistry Research,* 40 (12) (2001) 2679–2684.

128. C. Charcosset, I. Limayem, H. Fessi, The membrane emulsification process—A review. *Journal of Chemical Technology and Biotechnology,* 79 (2004) 209–218.

129. K. Hylton, S. Mitra, A microfluidic hollow fiber membrane extractor for arsenic (V) detection. *Analytica Chimica Acta,* 607 (1) (2008) 45–49.

130. K. Hylton, S. Mitra, Automated, on-line membrane extraction. *Journal of Chromatography A,* 1152 (2007) 199–214.

6

Low-Cost Tubular Ceramic
Support Membranes

6.1 Introduction

Catalytic membrane reactors that pursued two distinct functions—that is, reaction and separation—were prepared and broadly used for several applications. The performance of a catalytic membrane was related to its support configuration and surface texture, which was quite challenging and expensive with respect to raw materials [1,2]. Therefore, much attention had been paid to planning a novel route using low-cost raw materials for the fabrication of support of catalytic membranes and membrane reactors.

The cost of the precursors such as α-alumina, γ-alumina, zirconia, titania, and silica generally used in ceramic processing is significantly high and therefore contributes to the operating cost of membrane modules for industrial purposes. In the recent past, to overcome the issue of membrane cost, research in the fabrication of ceramic membranes has been focused toward the utilization of low-cost raw materials like natural raw clay [3–7], fly ash [8–11], apatite powder [12], and kaolin [7,13–19] for different applications.

6.2 Practical Example I: Fabrication of Low-Cost Tubular Ceramic Support

Tubular support membranes are used widely in catalytic membrane reactors for their ability to handle process streams with highly viscous solids and to minimize fouling by mechanical cleaning due to their small area per unit volume. Herein, the fabrication of low-cost tubular ceramic membrane via a dry compaction method using kaolin, feldspar, sodium metallicate (SM), and boric acid (BA) as the source materials and sawdust as pore-former is discussed as an example. The support membrane was prepared by thorough mixing and grinding of sawdust (sieved through a gyratory sieve shaker) along with dry inorganic raw materials of different weight ratios (Table 6.1) using a mixer machine and ceramic mortar (namely, SM1 to SM6). Kaolin was used to provide low plasticity and high refractory properties to the membrane support. Mechanical and thermal stability were contributed to the membrane by the addition of quartz. Sawdust, a lignocellulosic material, contains cellulose, hemicellulose, and lignin compounds. Sawdust is unique as a pore-former for the fabrication of ceramic membrane because it easily forms pores by removing cellulose, hemicellulose, and lignin compounds during heating and is cheap. The released lignocellulosic material emits via some uneven paths, which makes the membrane porous and contributes to the membrane porosity. Feldspar was utilized to obtain a glassy matrix in the early stage of the firing process. The compositions were selected on a trial basis to get the best results in terms of morphological properties, chemical and mechanical stability, and preparation cost. The ground mixture was placed into a cylindrical mold (outer diameter: 50×10^{-3} m; height: 50×10^{-3} m; thickness: 10×10^{-3} m) (Figure 6.1) and 9.81 MPa pressure applied by hydraulic press for 1 min to prevent propagation deformation and keep homogeneity in the raw material mixture; then it is sintered to obtain a rigid ceramic membrane. This investigation had two significant points that made this work novel. First, sawdust was adopted as a pore-former for the first time for the support of a tubular catalytic membrane reactor. Second, the combination of sawdust and kaolin was limited among the conventional kaolin-based membranes.

Wood, the cheapest lignocellulosic material, was the source of the inexpensive wood precursor sawdust [20,21]. Wood has previously been used as a natural composite material consisting of cellulose, hemicellulose, and lignin and it forms a cellular microstructure of high porosity, good strength, stiffness, and toughness [22,23]. Sawdust can be used in a variety of applications, including filters, ceramic bricks, catalytic membrane reactor support, etc. under a controlled manufacturing cost. The advantage of using sawdust over conventional pore-formers is the achievement of high porous structure economically.

TABLE 6.1

Composition of Raw Materials Used for Preparation of Ceramic Support Membrane

Name of Membrane Support	Composition Dry Basis (wt%)				Operating Pressure (MPa)
	Kaolin	Quartz	Sawdust	Feldspar	
SM1	40	20	30	10	
SM2	40	30	10	20	
SM3	30	10	20	40	9.81
SM4	40	25	10	25	
SM5	50	25	25	0	
SM6	50	0	25	25	

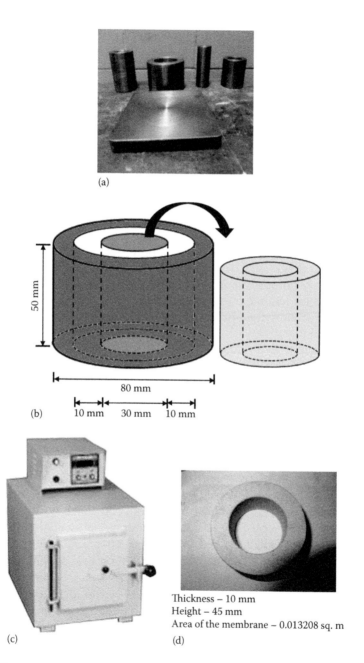

(a)

50 mm

80 mm

(b)   10 mm   30 mm   10 mm

Thickness – 10 mm
Height – 45 mm
Area of the membrane – 0.013208 sq. m

(c)   (d)

FIGURE 6.1
Fabrication of support membrane: (a) fabricated mold made of mild steel; (b) schematic diagram of the prepared tubular membrane support; (c) muffle furnace; (d) top view of the fabricated tubular ceramic membrane.

So far, sawdust has been used only in ceramic brick and cement manufacturing. However, there has been no report on the use of sawdust as a poreformer in ceramic membrane fabrication and its behavior—especially the change of physical properties and control on the morphology of membranes during sintering. In this case study on the fabrication of low-cost tubular ceramic support membrane, sawdust was used as a pore-former to reduce the manufacturing cost of the membrane [23–25], considering the role of the sawdust in the membrane fabrication.

6.2.1 Introduction of Sawdust as Pore-Former

6.2.1.1 Selection and Treatment of Sawdust

Selection of a pore-former with an appropriate particle size is the utmost task of a porous ceramic membrane. The pore size and porosity are important aspects in this regard. A pore-former with high particle size is necessary for a highly porous ceramic body. Selection of sawdust as a pore-former was attempted for ceramic membrane fabrication to achieve a better efficiency with a lower economy than those obtained by other standard pore-formers.

- Five different size ranges of sawdust screened through 30, 44, 60, 72, and 100 B.S.S. were considered to select the appropriate size of sawdust particle suitable for ceramic support membrane fabrication.
- Thermal modification of sawdust was performed to realize the change in physical properties of sawdust, which is the significant factor for understanding the pathway of making pores in ceramic membranes. The raw sawdust particles underwent three heat treatment steps. Firstly, the raw sawdust was dried at room temperature (28°C ± 2°C) for 24 h. After that, it was dehydrated at 100°C for 12 h in a muffle furnace followed by heating at 250°C for 24 h. Secondly, it was sintered from 250°C to the desired 850°C at a heating rate of 2°C/min for 5 h on the basis of thermogravimetric analysis (TGA). Then the thermally modified sawdust was cooled by an atmospheric cooling process implemented by switching off the furnace.
- The raw sawdust and burnt samples were separately dipped into acid (concentrated hydrochloric acid, pH 2) and alkali (NaOH, pH 12) solutions for 7 days to confirm any change in composition and surface structure.

6.2.1.2 Effect of Particle Size of Sawdust on Membrane Porosity and Pore Size

Particle size distribution of both raw sawdust (screened through B.S.S. 30, 44, 60, 72, and 100 mesh) and the sample mixtures for membrane fabrication are shown in Figure 6.2(a–f), respectively.

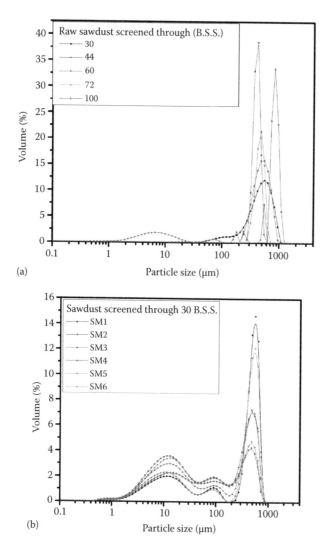

FIGURE 6.2
Volume weighed particle size distribution of (a) raw sawdust particles and sample mixtures sieved through mesh sizes of (b) 30 B.S.S., *(Continued)*

It is observed that particle size of sawdust screened through 30, 44, and 100 B.S.S. mesh provides wider particle size distribution, which can offer membranes with wide ranges of pores suitable for fabrication of membrane support of a catalytic membrane reactor. The average pore diameter of the fabricated membranes is obtained from different sample mixtures. To optimize suitable sawdust particle for the manufacturing of ceramic membrane, different weight ratios of raw materials with three different particle

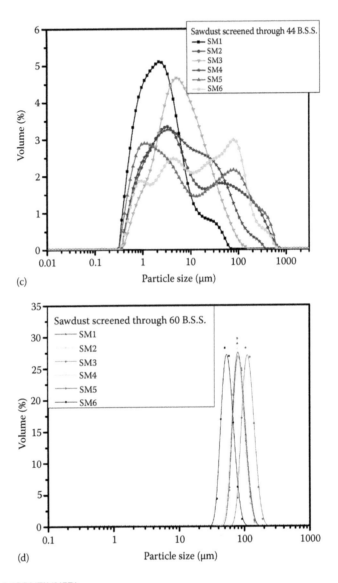

FIGURE 6.2 (CONTINUED)
Volume weighed particle size distribution of (c) 44 B.S.S., (d) 60 B.S.S.   *(Continued)*

size ranges of sawdust of 500, 355, and 150 μm, sieved through 30, 44, and 100 B.S.S. For the sample mixtures containing sawdust of differently sized particles sieved through 100 B.S.S. mesh, the average pore diameter varied between 0.05×10^{-9} and 0.10×10^{-9} m. Similarly, for membranes made using sawdust of differently sized particles sieved through 30 and 44 B.S.S. mesh, the average pore diameter varied between 0.04–0.10 μm and 0.06×10^{-9}–0.13 ×

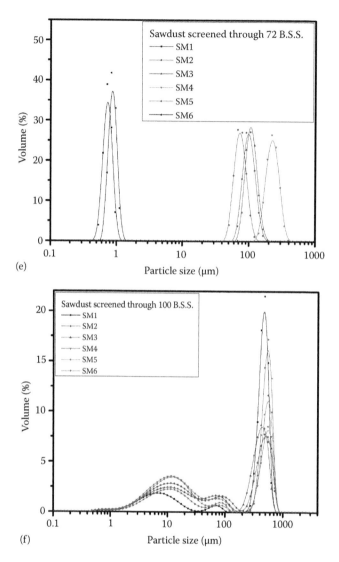

FIGURE 6.2 (CONTINUED)
Volume weighed particle size distribution of (e) 72 B.S.S., and (f) 100 B.S.S.

10^{-9} m, respectively. It was observed that the average pore diameters of all the fired membranes were very close to each other, unlike the particle size of sample mixtures together with different sizes of sawdust and its distribution. It is quite interesting to note that SM5 membrane containing sawdust screened through 100 and 30 B.S.S. mesh became cracked and damaged during sintering owing to large particle size and absence of feldspar. This result led to choosing 44 B.S.S. screened sawdust as pore-former for the fabrication of ceramic membrane support for catalytic membranes.

The porosity of the membrane fabricated using sawdust sieved through a 44 B.S.S. screen at 850°C was obtained at 34% by gas permeation study, indicating a highly porous support membrane, and used for the fabrication of catalytic membrane and membrane reactor.

6.2.1.3 Confirmation of Formation of Pores by Sawdust

Sawdust is unique as a pore-former for the fabrication of ceramic membrane because it easily forms pores by removing cellulose, hemicellulose, and lignin compounds in the form of carbon dioxide during heating. The path taken by the released CO_2 gas thereby creates the porous texture of the inorganic membrane and contributes to the membrane porosity. This phenomenon can easily be understood by identifying the functional group present in sawdust before and after sintering.

The peaks associated to the presence of hemicellulose, cellulose, and lignin in the raw sawdust were clearly visible in the fingerprint region between 1100 and 1800 cm^{-1} (Figure 6.3a). The absence of any band associated to cellulose, hemicellulose, and lignin in the fingerprint region was observed due to complete decomposition of cellulose, hemicellulose, and lignin, as shown in Figure 6.3(b). This phenomenon confirms formation of pores in the membrane.

6.2.2 Optimization Study

Performance of the membrane and its durability greatly depend on its morphology and mechanical strength. Therefore, membrane morphological properties such as porosity, pore size distribution, and mechanical properties like flexural strength are important parameters in ceramic processing. These parameters mainly depend on binder content, preparation pressure, sintering temperature, etc. Hence, it is essential to control these parameters for achieving a better performance membrane as a function of morphology and mechanical strength.

Nowadays, mathematical modeling and optimization techniques are frequently used in the field of membrane technology; statistical tools are especially well known for improving operating parameters and performance for membrane applications. In this study, design of expert software (DOE-RSM) was used to develop central composite design (CCD) with a full polynomial. The CCD procedure is an effective design tool that works for experimental design analysis with lack of fit testing that compares the variation around the model with "pure" variation within simulated observations and measures the adequacy of the quadratic response surface model, process optimization, development of second-order response models, estimating experimental error variance, etc. This tool follows three steps to conduct the experimental design: (1) statistical design of experiments, (2) assessment of coefficients through a mathematical model with the prediction of response, and (3) analysis of the

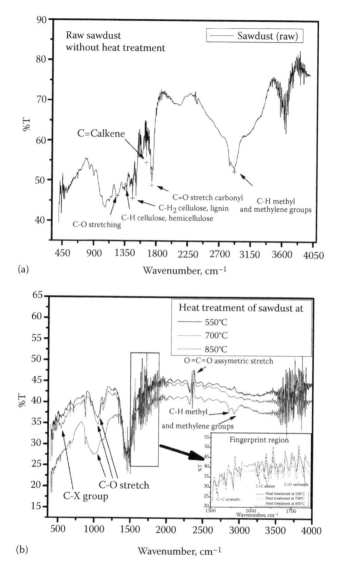

FIGURE 6.3
FTIR patterns of 44 B.S.S. mesh screened raw sawdust (a) before and (b) after sintering.

model's applicability. All the experimental response models were produced with respect to the input variables (experimental data). The experimental response models determined the flexural strength and porosity of the membrane, thus indicating the role of binders in membrane fabrication. The three input factors were chosen for designing purposes; their values were selected based on preliminary study and varied over five levels: the high value (+1), the center points (0), the low level (−1), and two outer points (−α and +α values).

The CCD contained design points and axial points that accompanied a total of 20 runs of the experiment and were used to study the data obtained from the experimental runs. These data were then used to optimize the amount of binders and preparation pressure. The response variables (i.e., flexural strength and porosity of membrane) were measured using a mathematical modeling selected from the CCD with significant terms, and the model was not aliased [24].

6.2.2.1 Raw Materials Content

Content of raw materials plays a significant role in the manufacturing of ceramic membranes; thus, optimization of raw material content is required. The optimization of raw materials excluding amount of binder is achieved on the basis of morphological characteristics such as pore size, porosity, and manufacturing cost. Membrane SM6 (kaolin: 50%; feldspar: 25%; sawdust: 25%) was chosen as the optimized composition for further study.

6.2.2.2 Sintering Temperature

The major weight loss of sawdust is noted in Figure 6.4 due to the removal of moisture and vaporization of organic matter (cellulose, hemicellulose, and lignin), which involves exothermic as well as endothermic reactions. This process of decomposition of sawdust consists of four major stages: (1) moisture removal, (2) hemicellulose decomposition, (3) cellulose decomposition, and (4) lignin decomposition, shown in Figure 6.4. The first downward slope

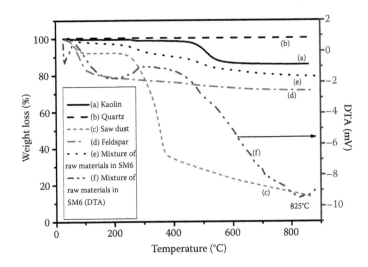

FIGURE 6.4
TGA/DTA plot of four individual raw materials and SM6 decomposed in presence of nitrogen atmosphere at 10°C/min.

at 25°C in the TGA curve of sawdust indicates 9% weight loss due to the removal of moisture. The second and third major weight losses (~55%) are noted from 210°C to 350°C due to decomposition of hemicellulose, cellulose, and the predehydration process of kaolin. The fourth stage—that is, lignin vaporization (weight loss ~ 16%)—also starts from 210°C and ends at 825°C; it is the main region of vaporization of lignocellulosic material. This phenomenon is confirmed by a small exothermic peak at 825°C displayed in the differential thermal analysis curve. The TGA curve of kaolin exhibits almost 10% weight loss due to the loss of the structural hydroxyl group from 425°C to 550°C, due to the transformation of kaolinite to metakaolinite. This study confirmed the temperature zones where moisture, cellulose, hemicellulose, and lignin burn completely. Based on this study, heating temperature of raw sawdust as well as optimum sintering temperature for membrane fabrication have been decided.

6.2.2.3 Binder Contents

Binder content is another key parameter that plays a most significant role in membrane fabrication, controlling membrane morphology and strength. In the previous section, we observed that the flexural strength of SM6 membranes is only 2 MPa at 850°C using the ASTM standard test (ASTM D790) without any binder. The number of contacts between the particles as well as the specific surface is reduced due to higher particle size of the SM6 membrane and absence of binder; thus, it is necessary to add binder. To investigate the effect of addition of binder on the membrane morphology and mechanical strength, a design plan was introduced that included three input factors: preparation pressure, amount of SM, and amount of BA (in weight percent). Outputs for each experiment were flexural strength and porosity, as shown in Table 6.2.

Maximum membrane porosity of 28% was obtained at 5% of binder content each, whereas the minimum porosity of 10% was attained at high binder content (10% each) (Figure 6.5). At high binder content, particles have a tendency to remain close to each other, which causes a decrease in porosity.

It was reported that high flexural strength was obtained at high preparation pressure with a larger amount of binders (see Figure 6.6a–c). As SM and BA form silicate and metallic metaborates at higher temperature during the sintering process, increases in particle–particle interconnectivity occurred.

On the other hand, among two binders, BA showed a superior influence to that of SM to obtain high flexural strength (Figure 6.6c). Increase in particle–particle interconnection reduced the voids between membrane layers and thus higher flexural strength was detected. Therefore, high pressure and high BA content are desirable conditions for producing a membrane with high flexural strength. After optimization study, optimized binder content was found as 7.50 wt% for each SM and BA.

TABLE 6.2

Design Arrangement and Experimental Responses for Central Composite Design (CCD)

Run	Factor 1: A, Preparation Pressure (MPa)	Factor 2: B, Amount of SM (%)	Factor 3: C, Amount of BA (%)	Response 1: R1, Flexural Strength (MPa)	Response 2: R2, Porosity (%)
1	7.84	5.00	10.00	10.54	15
2	9.81	7.50	4.21	8.35	24
3	11.77	5.00	5.00	9.72	26
4	9.81	7.50	7.50	11.68	21
5	9.81	4.21	7.50	11.53	22
6	9.81	7.50	7.50	11.68	21
7	11.77	10.00	5.00	11.82	18
8	9.81	7.50	7.50	11.68	21
9	12.39	7.50	7.50	11.99	16
10	7.84	10.00	10.00	11.29	13
11	7.23	7.50	7.50	5.32	24
12	9.81	7.50	10.79	12.33	11
13	9.81	7.50	7.50	11.68	21
14	7.84	10.00	5.00	5.01	25
15	9.81	10.79	7.50	11.85	13
16	9.81	7.50	7.50	11.68	21
17	9.81	7.50	7.50	11.68	21
18	7.84	5.00	5.00	4.46	28
19	11.77	10.00	10.00	12.82	10
20	11.77	5.00	10.00	13.78	13

6.3 Study of Phase Transformation and Microstructure of the Optimized Membrane

The XRD patterns of the ceramic membrane using optimized binder contents obtained from response surface methodology study—namely, C1 (SM content = 7.50%; BA content = 7.50%), along with C2 sintered at 550°C, 700°C, and 850°C—are shown in Figure 6.7. The reason behind the selection of three different sintering temperatures was to verify the change in morphology of the optimized ceramic membrane and phase transformation with sintering temperature. The additional composition—that is, C2 (SM content = 5%; BA content = 5%)—was introduced to verify the phase transformation with the change in binder contents, keeping the preparation pressure constant. C2 was introduced since it is difficult for the optimized membrane to identify any phase transformation with the variation in sintering temperature. The XRD diffractrogram clearly revealed that inoyite and nephiline are the predominant crystalline component

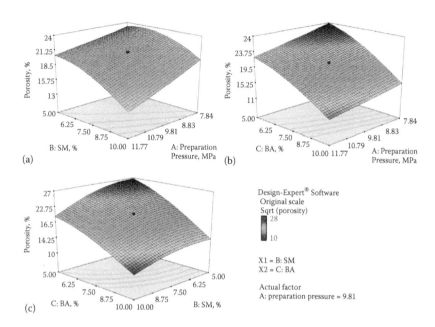

FIGURE 6.5
Response surface plotted on (a) SM content and preparation pressure, (b) BA content and preparation pressure, and (c) SM and BA content for membrane porosity.

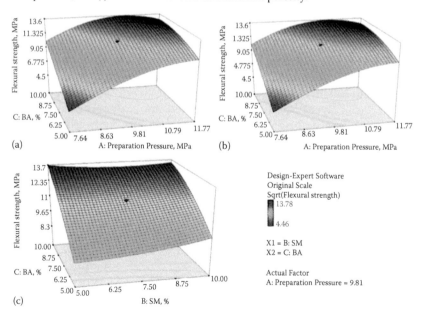

FIGURE 6.6
Response surface plotted on (a) SM content and preparation pressure, (b) BA content and preparation pressure, and (c) SM and BA content for membrane flexural strength.

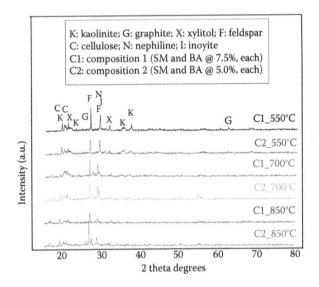

FIGURE 6.7
XRD patterns of the ceramic membrane (C1: optimized composition; C2: composition for validation) sintered at 550°C, 750°C, and 850°C with the formation of inoyite (PDF-00-006-0361) and nephiline (00-019-1176).

in the membrane matrix that signify membrane hardening with high crystallinity; this is in good agreement with the literature [24].

The microstructure of the ceramic matrix using field emission scanning electron microscopy (FESEM) images is shown in Figure 6.8. The microstructure of the ceramic matrix mainly depends on two factors: one is sintering temperature and the other is binder content. At lower temperature (550°C), a muddy and patchy matrix (Figure 6.8a) is observed, which corresponds to an interfacial layer of partially decomposed raw materials (mainly hemicellulose, cellulose, and lignin present in sawdust) during heating. A rough and pellet-like surface texture (Figure 6.8b) is detected when the membrane is sintered at 700°C, because of the increase in the rate of decomposition of

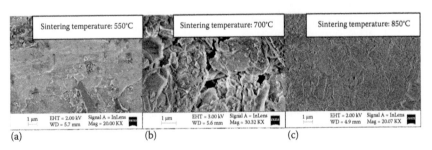

FIGURE 6.8
FESEM images of the ceramic membrane surface sintered at (a) 550, (b) 700, and (c) 850°C using optimized composition from CCD analysis.

lignocellulosic material with increasing temperature. On the other hand, at 850°C, the densification of the ceramic matrix increases due to complete decomposition of raw materials and binder burnout. Hence, the image shows highly consolidated and well-bonded matrix grains (Figure 6.8c).

6.3.1 Morphological Study of the Optimized Membrane

6.3.1.1 Archimedes' Test

According to the volumetric porosity method (Archimedes' test), the optimized membrane possessed 21% porosity and was slightly higher compared to the values calculated from gas permeation tests. This is because water remains intact in the pore interiors and pore walls because of the swelling nature of kaolin-based membrane.

6.3.1.2 Gas Permeation Test

The gas permeation experiment was implemented for the membrane prepared on the basis of optimized composition to determine the average pore diameter and the porosity. The gas permeance and the porosity of the optimized membrane sintered at 850°C was determined as 1.18×10^{-1} m^3 m^{-2} min^{-1} kPa^{-1} and 15%, respectively. The average pore diameter of the membrane was found to be 0.21 µm.

6.3.1.3 FESEM Image Analysis

Microstructure of the membranes can be affected by the sintering profile. It has been seen that higher temperature provides a much denser and smoother surface and also larger grain size and tiny pores. FESEM was used to observe the microstructure of the fabricated ceramic during the sintering process. Figure 6.8 displays FESEM images and the consistent grain size of sawdust-based tubular ceramic membrane after sintering at three different temperatures. With increasing sintering temperature, the grain size was increased and the pores started to be eliminated. At 850°C, most of the pores had vanished and a smooth structure was formed with complete disappearance of grain boundaries, as shown in Figure 6.8(c).

6.3.2 Mechanical Stability

The three-point bend test (ASTM D790) is performed to measure the flexural strength of all prepared membranes using a universal tensile test machine, applying a 20 MPa load, and maintaining a support span of 85 mm at a stroke rate of 0.5 mm.min^{-1}. Rectangular bar-shaped specimens are prepared at thickness/length ratio of 1:16 for determining the flexural strength at room temperature (28°C ± 2°C). The optimization study showed the desirable preparation pressure to be 9.81 MPa and the weight percentage of both

SM and BA to be 7.5% of total weight of the sample mixture, which gave optimized porosity of 21% and flexural strength of 11.55 MPa.

6.3.3 Chemical Stability

Raw sawdust and burned samples were separately dipped into acid (concentrated hydrochloric acid; pH 2) and alkali (NaOH; pH 12) solutions for 7 days separately to confirm any change in composition and surface structure. Chemical stability of the fabricated ceramic membrane was also verified by immersing the membrane into the acid (concentrated HCl; pH 2) and alkali (NaOH; pH 12) solution for 7 days. Firstl the membrane was weighed in a dry condition and dipped into the solutl of acid and alkali for 1 week at atmospheric conditions. Second, the membrane was weighed and the porosity estimated following the volumetric porosity determination technique. EDX analysis of the membranes before and after the corrosion test was performed to verify the change in elemental composition.

An increase in membrane porosity in the range of 9%–10% was observed in both acid and alkali media for the optimized membranes sintered at 850°C. However, a 3% decrease in porosity was detected for the membrane sintered at 550°C and 700°C during acid testing because of the production of silicon dioxide while SM reacted with HCl. A decrease in porosity in the range of 3%–8% was noted in alkali testing because of the production of SM while SM reacted with NaOH solution.

Energy-dispersive x-ray (EDX) studies of the membranes obtained from optimization study before and after the chemical stability test are shown in Table 6.3. It can be confirmed from the table that there is no significant change in elemental composition of the membrane in both acid–alkali media. Therefore, the observed trends in weight loss and EDX analysis during acid–alkali tests convey that there is no major change in elemental composition due to the presence of binders in ceramic processing.

TABLE 6.3

EDX Analysis of Optimized Membrane Materials Sintered at 850°C before and after Acid–Alkali Test

Components	Membrane (wt%)	Membrane after Acid Test (wt%)	Membrane after Alkali Test (wt%)
Oxygen	40.68	39.87	40.82
Silicon	27.12	27.91	27.28
Aluminum	29.50	30.18	29.66
Potassium	1.69	1.26	0.85
Sodium	1.01	0.78	1.39

6.3.4 Manufacturing Cost of the Fabricated Support Membrane

The economic feasibility assessment of the membrane was carried out on the basis of a few significant parameters, such as raw materials cost, energy consumption, and membrane mold manufacturing cost. The projected cost of the membrane was evaluated and is presented in Table 6.4. The estimated cost of the optimized membrane is approximately \$302/m². The total cost, including membrane mold preparation and other miscellaneous expenditures (50% of the estimated cost), is \$332/m².

The optimized membrane is less expensive than an α-alumina-based ceramic tube costing \$500–\$1000/m² now available in the market. This study proves that the ceramic membrane prepared by kaolin, sawdust, and feldspar, along with binders, would be attractive and more useful than polymeric membranes as well as ceramic tubular membranes installed in industrial arrangements. The cost of the membranes may vary with their performance and stability in different process applications. For example, in a recent study, two low-cost ceramic membranes were prepared by sintering at different temperatures using clay only (membrane A) and clay with small amounts of sodium carbonate, SM, and BA (membrane B) for the removal of chromate from aqueous solutions. The reported optimal compositions (based on wet basis) for membranes A and B were clay (70% w/w), water (30% w/w) and clay (70% w/w), sodium carbonate (3% w/w), SM (1.5% w/w), BA (1.5% w/w), and water (24% w/w), respectively. Overall performance of membrane B was better than that of membrane A. The average pore size, porosity, pore density, and flexural strength of membrane B sintered at 1000°C were determined to be 4.58 μm, 0.42, 2.06×10^{10} m^{-2}, and 11.55 MPa, respectively. Based on raw material prices, the membrane cost was estimated to be \$19/m² [6]. In a similar approach, a low-cost hydrophilic ceramic–polymeric composite membrane was synthesized for treatment of oily wastewater. Optimal compositions were provided (based on both dry bases) of clay (30% w/w), kaolin (30% w/w), sodium carbonate (5% w/w), SM (2.5% w/w), and BA (2.5% w/w) and (based on both dry bases) of clay (46.15% w/w), kaolin (23.08% w/w),

TABLE 6.4

Cost Assessment of Optimized Membrane

Raw Material Used	Weight (%)	Unit Price ($)	Energy Consumption at $0.08/kWh	Total Price ($/m²)
Kaolin	50	2.58		
Feldspar	25	1.13		
Sawdust	10	0	6.19	301.95
SM	7.5	0.08		
BA	7.5	0.08		

sodium carbonate (3.85% w/w), SM (1.92% w/w), BA (1.92% w/w), and water (23.08% w/w). The membrane cost was estimated to be \$33.42/m^2, calculated on the basis of prices of the chemicals; the cost can reach \$100/m^2, taking the cost of manufacturing and shipment into account [7].

6.4 Practical Example II: Fabrication of Low-Cost Ceramic Support

In order to obtain low-cost ceramic membranes by minimizing the cost of the firing step, another study was conducted recently to determine the possibility of preparing porous chamottes (a calcined clay) to be used in a low-cost membrane's synthesis as an alternative to organic pore-forming agents [26].

The clay UA-50 was selected to prepare the chamottes. The reported clay composition was approximately 65.6% SiO_2, 22.8% Al_2O_3, 0.6% Na_2O, 2.3% K_2O, 1.3% TiO_2, 1.1% Fe_2O_3, 0.3% CaO, and 0.5% MgO, with a loss on ignition of 6.5 wt%. The chamottes were obtained from a mixture of 90 wt% clay and 10 wt% starch, using three different starches:

- S1 (potato starch; Roquette Freres S.A., Lestrem, France)
- S2 (pea fiber L50M; Roquette Freres S.A.)
- S3 (soluble potato starch Pregeflo P100; Roquette Freres S.A.)

Several chamottes (10S1-D, 10S2-D, 10S3-D, 10S2-W1, 10S1-W2, 10S3-W2, 30S1-D, 30S1-W2) were prepared by adding different amounts of starch (10%–30%). Different preparation methods were used and the clay–starch agglomerates were fired in an electric kiln to peak temperatures from 1050°C to 1200°C with a heating rate of 10°C/min and a soaking time of 1 h.

Ceramic membranes were prepared with the chamottes and a mixture of clay UA-50, micronized sodium feldspar, and feldspathic sand (AFS-125). Different weight percentages of chamotte ranging from 15 wt% to 60 wt% were added to the clay–feldspar mixture, whose composition was approximately 72% SiO_2, 17.6% Al_2O_3, 4.2% Na_2O, 1.5% K_2O, 0.6% TiO_2, 0.5% Fe_2O_3, 0.3% CaO, and 0.2% MgO with a loss on ignition of 2.9 wt%.

The membrane compositions were moistened to a water content of 0.055 kg water/dry solid kg. Disk-shaped test specimens of 50 mm diameter and 3–4 mm thickness were formed by uniaxial dry pressing at 300 kg/cm^2 and dried in an oven at 110°C. The green specimens were then fired in a fast electric kiln at different peak temperatures, ranging from 1050°C to 1125°C. The heating rate was maintained at 25°C/min, with a holding time of 60 min at peak temperature.

The particle size distribution of the starches and the characteristic diameters D_{10}, D_{50}, D_{90}, D_V, and D_S were calculated by dry laser diffraction. The humidity was determined from the weight loss after drying at 110°C in an electrical oven (it was given as kilogram of water by 100 kg of dry solid) and the ash content was determined by calcining every starch at 1000°C. The particle size distribution of the chamottes and the membranes was measured by mercury intrusion porosimetry. Surface area of the chamottes was determined using the Brunauer–Emmett–Teller method. Apparent porosity, measured as water uptake according to standard UNE-EN ISO 10545-3, was also determined and the permeability coefficient for water obtained with a liquid permeameter (LEP101-A, PMI, Ithaca, NY). Additionally, the microstructure of the chamottes as well as the membranes was examined by field emission gun scanning electron microscopy.

After the experiments, the authors concluded that

- The chamottes consisted of hard, porous agglomerates with an interconnected pore network and allowed membranes with considerably shorter firing cycles to be made than when starch was used as the pore-former.

- Permeability values increased with firing temperature up to 1075°C and then decreased due to the membrane's densification, which reduces the interconnectivity of the pore network. As a result, 1075°C has been set as optimum firing temperature of the membranes, though the maximum mechanical strength of the membrane is obtained at 1200°C.

- Permeability of the membranes increased with chamotte content and, for values lower than 45 wt%, there were no clear differences among the different chamottes. However, with 60 wt% of chamotte, the highest permeability corresponded to the membrane prepared with the 10S1-W2 chamotte.

- Membranes with more than 60 wt% chamotte could not be obtained due to processing problems.

- Low-cost membranes with narrow particle size distribution (around 2 μm) can be successfully obtained with short firing cycles when porous chamottes are used instead of traditional pore-formers.

6.5 Comparison to High-Cost Ceramic Membranes

The costs of the precursors such as α-alumina, γ-alumina, zirconia, titania, and silica generally used in ceramic processing are significantly high and

TABLE 6.5

Comparisons of Manufacturing Cost of Various Studies

Materials	Parameters Considered	Configuration	Cost	Ref.
Kaolin, quartz, and calcium carbonate; sodium carbonate; BA; SM; water	Only raw materials cost	Disk	$130/m²	[15]
Kaolin, quartz, and calcium carbonate		Circular disk	$81/m²	[16]
Kaolin, quartz, and calcium carbonate; sodium carbonate; BA; SM; polyvinyl alcohol		Circular disk	$78/m²	[19]
Kaolin, quartz, and calcium carbonate; sodium carbonate; BA; SM; water		Tubular	$110–$135/m²	[28]
Kaolin, sawdust, feldspar, BA, SM	Raw materials cost + mold preparation cost + energy consumption cost + miscellaneous	Tubular	$332/m²	[24]

therefore contribute to the high operating cost of membrane modules for industrial purposes. In the recent past, to overcome the issue of membrane cost, research in the fabrication of ceramic membranes has been focused on the utilization of cheaper raw materials, such as natural raw clay, fly ash, apatite powder, dolomite, and kaolin, along with pore-formers such as calcium carbonate and sodium carbonate [3–8,12–19,27,28]. Manufacturing costs of the fabricated ceramic membranes (disk and tubular shaped) reported in the previously published literature are given in Table 6.5.

Thus far, the previous discussion has indicated no energy consumption cost, and other costs, such as that for mold preparation, were not considered during cost evaluation of the ceramic membranes. On the other hand, the present study has already proven successful use of sawdust as a pore-former and its economic and strategic value to lower temperature, which provides excellent membrane morphology, strength, and chemical sustainability.

6.6 Summary

In previous chapters, we have discussed information about and several benefits of low-cost ceramic membranes broadly. In this chapter, we have described two practical examples of low-cost ceramic membranes of different shapes for different applications. To summarize from the examples:

- Raw materials play a significant role in minimizing manufacturing cost.
- Sawdust (consisting of cellulose, lignin, and hemicellulose), starch (a large number of glucose units), and chamottes (a calcined clay consisting of hard porous agglomerates with an interconnected pore network) have been successfully implemented in ceramic membrane fabrication.
- Ceramic membranes made of sawdust provide not only good morphological (21% porous), thermal, chemical (excellent resistance in highly acidic and alkaline media), and mechanical (11.55 MPa) properties, but also low manufacturing cost ($332/m^2).
- Ceramic membranes made of chamottes have also proved to have good morphological characteristics along with excellent mechanical strength.

Though the invention and uses of cheap raw materials are involved in fabrication of ceramic membranes extensively nowadays, many more inexpensive and waste products can also be used as raw materials in ceramic membrane fabrications. More research is needed for further developments.

References

1. H.W.J.P Neomagus, G. Saracco, A. Versteeg, Fixed bed barrier reactor with separate feed of reactants. *Chemical Engineering Communications,* 184 (2001) 49–69.
2. M.P. Pina, M. Menendez, J. Santamaria, The Knudsen-diffusion catalytic membrane reactor: An efficient contactor for the combustion of volatile organic compounds. *Applied Catalysis B: Environment,* 11 (1) (1996) L19–27.
3. N. Saffaj, M. Persin, S.A. Younssi, A. Albizane, M. Bouhria, H. Loukili, H. Dacha, A. Larbot, Removal of salts and dyes by low ZnAl$_2$O$_4$–TiO$_2$ ultrafiltration membrane deposited on support made from raw clay. *Separation and Purification Technology,* 47 (2005) 36–42.
4. N. Saffaj, M. Persin, S.A. Younsi, A. Albizane, M. Cretin, A. Larbot, Elaboration and characterization of microfiltration and ultrafiltration membranes deposited on raw support prepared from natural Moroccan clay: Application to filtration of solution containing dyes and salts. *Applied Clay Science,* 31 (2006) 110–119.
5. S. Khemakhem, A. Larbot, R.B. Amar, New ceramic microfiltration membranes from Tunisian natural materials: Application for the cuttlefish effluents treatment. *Ceramics International,* 35 (2009) 55–61.
6. S. Jana, M.K. Purkait, K. Mohanty, Preparation and characterization of low-cost ceramic microfiltration membranes for the removal of chromate from aqueous solutions. *Applied Clay Science,* 47 (2010) 317–324.

7. P. Mittal, S. Jana, K. Mohanty, Synthesis of low-cost hydrophilic ceramic–polymeric composite membrane for treatment of oily wastewater. *Desalination*, 282 (2011) 54–62.

8. Y. Dong, X. Liu, Q. Ma, G. Meng, Preparation of cordierite-based porous ceramic microfiltration membranes using waste fly ash as the main raw materials. *Journal of Membrane Science*, 285 (2006) 173–181.

9. Z. Wei, J. Hou, Z. Zhu, High-aluminum fly ash recycling for fabrication of cost-effective ceramic membrane supports. *Journal of Alloys and Compounds*, 683 (2016) 474–480.

10. J. Liu, Y. Dong, X. Dong, S. Hampshire, L. Zhu, Z. Zhu, L. Li, Feasible recycling of industrial waste coal fly ash for preparation of anorthite-cordierite based porous ceramic membrane supports with addition of dolomite. *Journal of the European Ceramic Society*, 36 (4) (2016) 1059–1071.

11. J. Cao, X. Dong, L. Li, Y. Dong, S. Hampshire, Recycling of waste fly ash for production of porous mullite ceramic membrane supports with increased porosity. *Journal of the European Ceramic Society*, 34 (13) (2014) 3181–3194.

12. S. Masmoudia, A. Larbot, H.E. Feki, R.B. Amara, Elaboration and characterization of apatite based mineral supports for microfiltration and ultrafiltration membranes. *Ceramics International*, 33 (2007) 337–344.

13. A. Potdar, A. Sukla, A. Kumar, Effect of gas phase modification of analcime zeolite composite membrane on separation of surfactant by ultrafiltration. *Journal of Membrane Science*, 210 (2002) 209–225.

14. C. Neelakandan, G. Pugazhenthi, A. Kumar, Preparation of NOx modified PMMA–EGDM composite membrane for the recovery of chromium (VI). *European Polymer Journal*, 39 (2003) 2383–2391.

15. B.K. Nandi, R. Uppaluri, M.K. Purkait, Preparation and characterization of low cost ceramic membranes for microfiltration applications. *Applied Clay Science*, 42 (1–2) (2008) 102–110.

16. D. Vasanth, G. Pugazhenthi, R. Uppaluri, Fabrication and properties of low cost ceramic microfiltration membranes for separation of oil and bacteria from its solution. *Journal of Membrane Science*, 379 (2011) 154–163.

17. V. Singh, M.K. Purkait, V.K. Chandaliya, P.P. Biswas, P.K. Banerjee, C. Das, Development of membrane based technology for the separation of coal from organic solvent. *Desalination*, 299 (2012) 123–128.

18. A. Agarwal, M. Pujari, R. Uppaluri, A. Verma, Preparation, optimization and characterization of low cost ceramics for the fabrication of dense nickel composite membranes. *Ceramics International*, 39(7) (2013) 7709–7716.

19. S. Emani, R. Uppaluri, M.K. Purkait, Preparation and characterization of low cost ceramic membranes for mosambi juice clarification. *Desalination*, 317 (2013) 32–40.

20. Wood handbook—Wood as an engineering material (general technical report, FPL-GTR-113), 1999, USDA, Forest Service, Madison, WI.

21. A.P. Scniewind, *Concise encyclopedia of wood and wood-based materials*, Pergamon Press, New York, 1989.

22. M.F. Ashby, K.E. Easterling, R. Harrysson, S.K. Maity, The fracture and toughness of woods. *Proceedings Royal Society London, A*, 398 (1985) 261–280.

23. S. Bose, C. Das, Preparation and characterization of low cost tubular ceramic support membranes using sawdust as a pore-former. *Materials Letters*, 110 (2013) 152–155.

24. S. Bose, C. Das, Role of binder and preparation pressure in tubular ceramic membrane processing: Design and optimization study using response surface methodology (RSM). *Industrial Engineering & Chemistry Research,* 53 (31) (2014) 12319–12329.
25. S. Bose, C. Das, Sawdust: From wood waste to pore-former in the fabrication of ceramic membrane. *Ceramics International,* 41 (3) part A (2015) 4070—4079.
26. M.–M. Lorente-Ayza, M.-J. Orts, A. Gozalbo, S. Mestre, Preparation of chamottes as a raw material for low-cost ceramic membranes. *International Journal of Applied Ceramic Technology,* 13 (6) (2016) 1149–1158.
27. N. Saffaj, S.A. Younssi, A. Albizan, A. Messouadi, M. Bouhria, M. Persin, M. Cretin, A. Larbot, Preparation and characterization of ultrafiltration membranes for toxic removal from wastewater. *Desalination,* 168 (2004) 259–263.
28. D. Ghosh, M.K. Sinha, M.K. Purkait, A comparative analysis of low-cost ceramic membrane preparation for effective fluoride removal using hybrid technique. *Desalination,* 327 (2013) 2–13.

7

Fabrication of a Low-Cost Tubular Catalytic Membrane Reactor

7.1 Introduction

Catalytic membranes and membrane reactors present numerous advantageous in comparison with conventional separation processes. Nevertheless, the main drawbacks are the manufacturing process and manufacturing cost due to the use of overpriced raw materials. Hence, it is necessary to combine a low-cost ceramic support membrane and catalyst for the fabrication of catalytic membrane. The fabrication procedure includes synthesis of catalyst and coating of the catalyst over the membrane surface with a solid lubricant (i.e., molybdenum disulfide, which can provide long life and no contamination and can be sustained in the harsh environment). Paint coating, a novel technique, is introduced for simplicity, ease of methodology, and minimization of cost. In this chapter, a complete description of experimentation involving the fabrication and characterization of the catalytic membrane reactor (CMR) is outlined. We have also discussed the viability and economic feasibility of the coating method and evaluated the morphological and structural behavior of the coated ceramic surface when it is lubricated with catalyst to assess the influence of lubricant on such coatings. The manufacturing cost of the CMR is also evaluated.

7.2 Synthesis of the Catalyst

7.2.1 Chemicals for Catalyst Preparation

Cobalt nitrate hexahydrate and ammonium molybdate tetrahydrate are used as Co and Mo precursors, respectively. Gamma-Al_2O_3 with high (100×10^{-3} $m^2.Kg^{-1}$) and low (60×10^{-3} $m^2.Kg^{-1}$) surface areas is used as catalyst support. Milli-Q water is used for slurry preparation.

7.2.2 Preparation of Catalyst

Mo-Co/γ-Al$_2$O$_3$ catalyst has been prepared via impregnation by mixing Mo and Co precursors and water for γ-alumina support, optimum Mo content, and calcination temperature. This method allows well-controlled distribution of the species. Accurately weighed 20 × 10^{-3} kg of catalyst (by maintaining the weight percentage) is prepared for this study. The support material is mixed with precursors and Milli-Q water, by maintaining a 1:1 water/solid ratio to get a slurry solution, and kept at the ambient temperature (28°C ± 2°C) for 30 min to retain the homogeneity of the sample. The sample is then placed in a muffle furnace and temperature is increased and kept at 120°C for 9 h, to remove excess water from the solution, and calcined at 400°C for 5 h at a heating rate of 5°C/min. The catalyst compositions of different loading of Mo and Co precursors with different synergistic ratio (0.18–0.50) are mentioned elsewhere [1].

7.2.3 Characterization Techniques

7.2.3.1 BET Surface Area Analysis

The specific surface area and pore size of the catalysts are characterized by a Brunauer–Emmett–Teller (BET) surface area analyzer using N$_2$ adsorption at liquid nitrogen temperature. The sample has been previously weighed (approx. 50 mg) and kept in a sample tube and then degassed for 1 h at 120°C before being exposed to the analysis conditions. The sample tube is then fixed at an analysis port, the bottom part of which is merged into a dewar containing liquid nitrogen. Specific surface area is obtained by a multipoint BET equation in the 0.05 to 1.0 P_s/P_0 range.

7.2.3.2 Particle Size Analysis

The catalyst particle diameter is measured by dynamic light scattering in a particle size analyzer by sonication of samples into water as dispersing agent in an ultrasonicator. The samples are irradiated with a He–Ne laser (wavelength = 632.8 nm). The intensity fluctuations of the scattered light are analyzed to obtain the correlation function, from which the size distribution is obtained by fitting a multiple exponential to it by a non-negative least-squares method.

7.2.3.3 Fourier Transform Infrared Spectroscopy

Fourier transform infrared spectroscopy (FTIR) spectra are recorded in transmission mode using a diffuse reflectance assembly over the range of 400 to 4000 cm^{-1} to realize the formation of a complex between cobalt–molybdenum precursor and alumina support. A background is first measured with dried potassium bromide (KBr). Then the catalyst sample of 1 to 2 mg is ground to a fine powder, mixed with the KBr using ceramic porcelain mortar, and then loaded into a sample holder mounted inside the instrument.

7.2.3.4 X-ray Diffraction

The change in phase of catalysts with the change in calcination temperature and dispersion of active phase is predicted by a diffractometer with a Bragg/Brentano geometry equipped with a proportional detector using Cu Kα radiation ($\lambda = 1.54060$ Å) of 40 kV and 40 mA in the 2θ range from 5° to 80° with 0.05° steps at a rate of 1 s per step. The various phases are predicted on the basis of the Joint Committee on Powder Diffraction Standards powder diffraction file cards.

7.2.3.5 Field Emission Scanning Electron Microscopy

The surface texture of the catalysts is analyzed by field emission scanning electron microscopy (FESEM) operating at 3 kV. A small amount of the catalyst sample is mounted on an aluminum stab with carbon tape, coated with gold particles under vacuum, and then mounted on the FESEM sample holder.

7.2.3.6 Electron Spin Resonance

Electron spin resonance (ESR) spectroscopy is a very powerful and sensitive method for characterization of the electronic structures of materials with unpaired electrons. The ESR spectra of the catalyst sample (~30 mg) are taken at room temperature with a spectrometer to verify the presence of a metal complex during catalyst preparation.

7.2.3.7 Temperature Programmed Reduction

The prepared catalysts are heated overnight in a hot air oven and cooled to room temperature. Temperature programmed reduction (TPR) in hydrogen atmosphere takes place until the baseline is stabilized. The flow rate of hydrogen is maintained at 30 cm³ min⁻¹. TPR is performed up to 900°C for 1 h, at 10°C min⁻¹, in 30 mL min⁻¹ of 50% H_2/N_2 mixture, followed by degassing of the sample (He purging) at 120°C, in 30 mL min⁻¹, for a half hour. The temperature is recorded using ChemiSoft Tpx V1.02 software.

7.2.3.8 CO Chemisorption

Metal dispersion, metallic surface area, and active particle diameter of the catalyst are measured in Micromeritics® Unit 1 (2720). The catalyst samples are first degassed at 150°C in helium atmosphere for 30 min at the degassing port using a sample tube. After cooling down to room temperature, the sample tube is transferred to the analysis port for the reduction analysis of the sample. The reduction procedure has been done up to 900°C for 1 h at a heating rate of 10°C min^{-1}. The chemisorption study is then analyzed by injecting CO gas until the reduction temperature cools down to room temperature under argon atmosphere. The formulae used for the calculation of metal dispersion and metallic surface area are as follows:

- CO chemisorption on 0.5 wt% Mo on alumina:
  - Volume of active gas injected from a syringe 10% CO/He:

$$V_{inj} = V_{syr} \times \frac{T_{std}}{T_{amb}} \times \frac{P_{std}}{P_{amb}} \times \frac{\%A}{100} \tag{7.1}$$

- Calculating the volume chemisorbed:

$$V_{ads} = \frac{V_{inj}}{m_s} \times \sum_{i=1}^{n} \left(1 - \frac{A_i}{A_f} \right) \tag{7.2}$$

- Percentage of metal dispersion (MD):

$$\%D = S_f \times \frac{V_{ads}}{V_g} \times \frac{m.w}{\%M} \times 100 \times 100 \tag{7.3}$$

- Active metal surface area (MSA) per gram of sample:

$$MS_{as} = S_f \times \frac{V_{ads}}{V_g} \times N_A \times \delta_m \times \frac{m_s^2}{10^{18} nm^2} \tag{7.4}$$

7.2.3.9 Laser Raman Spectroscopy

The local structure of the oxidic precursors is studied by laser Raman spectroscopy (LRS) using a spectrometer with 632.8 nm He–Ne, run in a back-scattered confocal arrangement. The samples are pressed in a microscope slide and the scanning time is set to 30 s. Raman spectra are recorded in air at room temperature with a resolution of 2 cm^{-1} in the scanning range of 100 to 1200 cm^{-1}.

7.2.3.10 Transmission Electron Microscope

For transmission electron microscope (TEM) analysis, a tiny amount of each sample is dispersed in ethanol and then sonicated for 30 min; a drop of the suspension is placed on a holey carbon thin film that is supported by a copper TEM grid. TEM images, selected-area electron diffraction (SAED) patterns, and chemical analyses of the materials are obtained using a 200 KV TEM equipped with an energy-dispersive x-ray (EDX) spectrometer.

The significant findings drawn from the study are summarized as follows:

- The catalyst prepared by depositing metal precursors via impregnation by soaking on a γ-Al_2O_3 (high surface area) surface was proved better than the γ-Al_2O_3 (low surface area) surface-based catalysts in terms of morphology, metal dispersion, and metallic surface area. No such influence of support material on catalyst textural properties (from FTIR and x-ray diffraction [XRD] analysis) was observed except physicochemical properties such as pore size, pore volume, and specific surface area. Alumina (high surface area) surface-based catalysts showed a wide range of pore sizes along with high MD and MSA due to negligible particle accumulation or pore restriction, which may be useful for solid-catalyzed reactions.

- More compact and agglomerated surface structures were observed in the FESEM images of all the catalyst samples containing higher Mo content (16 wt%). MD and MSA increased with increase in Mo content for the γ-alumina (high surface area) supported catalyst.

- Higher calcination temperature initiated cracks on the surface of the metal particles. As a result, MD and MSA decreased for the alumina (high surface area) surface-based catalyst.

- The ESR technique confirmed that the oxo–Mo^{5+} and -Mo^{6+} bind with alumina and that Mo^{4+} interacts with Co^{2+}. The study also concluded that metal dispersion increased with increase in metal content. The presence of crystalline MoO_3 and $CoMoO_4$ was established by SAED. TEM images showed a well-distributed dispersion of metal precursors over the support matrix.

- The optimum Mo content on alumina and the optimum calcination temperature were 16 wt% and 400°C, respectively, as shown by textural and physicochemical properties.

7.3 Coating of Catalyst on Support Membrane

After synthesis of the catalyst, it is coated over the outer surface of the tubular ceramic support membrane using a paint coating technique and four catalytic membranes are obtained—namely, M1, M2, M3, and M4. Pure molybdenum disulphide (MoS_2) and silica gel are purchased from M/s Sisco Research Laboratories Pvt. Ltd. (India). Sodium silicate powder (meta) ($Na_2SiO_3.9H_2O$)

is obtained from M/s Loba Chemie Pvt. Ltd. (India). These chemicals are used for the preparation of the lubricant solution. Sodium metasilicate is used as binding material to perform MoS_2 coating on the ceramic specimen. Silica gel is used as an additive to verify the change in membrane morphology. The weights of Na_2SiO_3 and the catalyst are taken as twice the weight of MoS_2 and the silica gel is added in the lubricant mixture by maintaining the ratio of 1:2:0.05 (MoS_2: Na_2SiO_3: silica gel).

The optimized tubular ceramic support membrane of 45×10^{-3} m height, 50×10^{-3} m diameter, and 10×10^{-3} m thickness (Figure 7.1a) has been taken for lubrication and fabrication of the catalytic membrane.

An optimum proportion maintaining one part of MoS_2 and two parts of sodium silicate/catalyst is experimentally obtained via three steps for the preparation of lubricant mixture. Firstly, 4.6 g of sodium metasilicate is added to 100 mL of Milli-Q water and the solution is mixed thoroughly. Secondly, the prepared solution is heated to 80°C for 1 h to evaporate the moisture present in the solution. Thirdly, the lubricant solution is applied four to five times on the support membrane by brush and then the specimen is heated in the hot air oven for 1 h at 150°C. Finally, a soft brush is used to remove loose MoS_2 powder from the surface of the support membrane after heating.

7.3.1 Study of Membrane Surface Morphology

The porosity, pore size, and surface topography of the fabricated and optimized catalytic membrane (M3) are determined by the volumetric porosity method. A portion of the specimen is cut into a small piece to observe the surface morphology of the specimen using FESEM, XRD, FTIR, and EDX mapping throughout the membrane surface to observe the surface morphology of the specimen. The average membrane pore size of all the samples is determined by ImageJ software (version 1.37). Fabricated and optimized

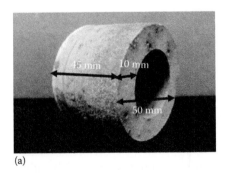

(a)   (b)

FIGURE 7.1
(a) Ecoceramic low-cost ceramic tubular membrane; height: 45×10^{-3} m, diameter: 50×10^{-3} m, and thickness: 10×10^{-3} m; (b) catalytic membrane after coating of catalyst using solid lubricant.

catalytic membrane shows highest porosity (i.e., 0.35). This phenomenon is due to the absence of binder in the support membrane. An absence of binder reduces the particle–particle interaction and creates higher voids, which trap water in the wall of the membrane, thus increasing the porosity. The average pore diameter of the optimized catalytic membrane is found as 1.76×10^{-6} m from FESEM images using ImageJ software.

7.3.2 Measurement of Film Thickness

The MoS_2 film thickness on the support membrane is determined by measuring mass and area of the coated specimen, which is called the gravimetric coating determination technique (ASTM standard B767-88). The specimen is weighed before and after coating to determine mass of the MoS_2 coating. The thickness of the membrane is determined using the following equation:

$$T_c = V_c/A \qquad (7.5)$$

where
  T_c is the thickness in m
  V_c is the volume of the coated surface $(= m/d_{sl})$ in m³
  m is the mass of the coating in kg
  d_{sl} is the density of solid-lubricant solution in Kg.m⁻³
  A is the area in m

The thicknesses of all the samples have been measured by the gravimetric coating determination technique and are in the range of 9×10^{-8} to 9.5×10^{-8} m.

7.3.3 Distribution of Catalyst over the Membrane Surface

The dispersion of the catalyst particles into the lubricant solution is confirmed by XRD peaks. EDX maps (Figure 7.2) clearly demonstrate that all the

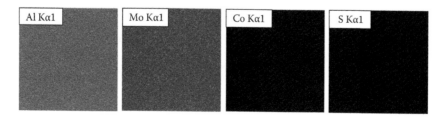

FIGURE 7.2
EDX mapping micrographs confirm fabrication of the catalytic membrane.

elements are distributed throughout the membrane. The uniform distribution of all the elements in EDX maps explains that the elements dispersed well into the membrane surface.

7.4 System Economics

In previous chapters, the economic feasibility of the membrane was determined on the basis of raw materials cost, energy consumption, membrane mold manufacturing cost, etc. The approximate manufacturing cost has been estimated as $332/m², including membrane mold preparation and other miscellaneous expenditures (10% of the estimated cost). Now, the cost analysis is performed including cost of catalyst preparation and coating solution. The probable cost of the optimized catalytic membrane is evaluated as $381/m², as presented in Table 7.1.

The projected cost is far less expensive than an α-alumina-based ceramic tube costing $500–$1000/m² available on the market [2]. This study proves that the approach adopted to coat catalyst on the surface of a support

TABLE 7.1

Estimated Cost of the Catalytic Membranes

Combination for Coating Solution	Cost in INR	Cost in $	Total in INR	Total in $
M1	2,761.18	44.13	23,708.58	380
			23,744.07	381
			23,708.58	380
M2	2,761.05	44.13	23,708.45	380
			23,743.94	381
			23,720.06	380
M3	2,758.23	44.26	23,705.63	380
			23,741.12	381
			23,717.24	380
M4	2,758.10	44.26	23,705.5	380
			23,740.99	381
			23,717.11	380
Combination of catalyst				
SC1	616.40	9.89	Approximate total cost is	
SC2	651.89	10.46	$381/m²	
SC3	628.01	10.08		
Support membrane	20,331.00	332.00		

Note: INR = Indian rupee rate.

membrane for the fabrication of the CMR would be attractive and more useful than the conventional preparation processes. The cost of the CMR may vary with performance and stability in different process applications.

7.5 Other Fabrication Methods

In this section, a few conventional fabrication methods of CMR will be discussed briefly.

Case 1: Five CoMo/γ-Al$_2$O$_3$ samples are prepared using the equilibrium deposition filtration (EDF) method, co-EDF, co-WET, WET (wet impregnation technique), and s-DRY (dry impregnation technique). Gamma-Al$_2$O$_3$ is used as the support. In the EDF sample, the molybdenum species are first mounted by EDF and then the cobalt species are deposited on the calcined Mo/γ-Al$_2$O$_3$ precursor solid by simple dry impregnation.

In the second sample both the molybdenum and the cobalt species are mounted on the surface from the same mixed Mo–Co solution using the co-EDF method. In the co-WET sample both Mo and Co species are mounted on the support surface by co-WET impregnation. In the WET sample the Mo species are deposited by wet impregnation and then the Co species are deposited on the calcined Mo/γ-Al$_2$O$_3$ precursor solid by dry impregnation. Lastly, in the fifth sample the Mo species are mounted on the γ-Al$_2$O$_3$ surface by successive dry impregnations, each time on the calcined sample of the previous step, and then the Co species are deposited on the Mo/γ-Al$_2$O$_3$ precursor solid by simple dry impregnation [3].

Case 2: In this study, a commercial cylindrical cordierite specimen (typical dimensions: diameter 1.5 cm and length 2 cm) has been used as support. In such honeycombs each channel has a square cross section with dimensions 1 × 1 mm, whereas the wall thickness is 200 mm.

The γ-Al$_2$O$_3$ layer deposition is carried out by dip-coating from the liquid phase using a sol prepared with a dispersible commercial colloidal pseudo-boehmite (γ-AlOOH) powder. Upon calcination in the temperature range of 500°C–700°C, this powder converts to the γ-Al$_2$O$_3$ phase. The solids content of the sol is kept constant at 10 wt% because, beyond this value, the sol viscosity increases significantly and the sol is not suitable for homogeneous coating. The sol pH is adjusted to 4.0 by the addition of HNO$_3$.

The cordierite specimens are then immersed in the sols for 1 min. In order to check for the reproducibility of results, three to five samples per case are examined. The loaded specimens are withdrawn, the sol remaining in the channels is allowed to drain, and removal of excess sol that formed menisci on the channel walls is achieved by blowing air through the honeycomb channels. After that, the coated honeycombs are dried at 110°C for 2 h and calcined at 600°C for 2 h so that a γ-Al_2O_3 coating following the substrate can be formed [4].

The other conventional techniques to fabricate CMRs have been detailed in Chapter 5.

7.6 Summary

A novel approach to fabricate a catalytic membrane and membrane reactor with an example has been presented in this chapter and compared with the other conventional techniques. The topic described in this chapter is summarized as follows:

- There are steps to prepare a catalytic membrane: fabrication of a ceramic support membrane, preparation of a catalyst, and coating of a catalyst over the exterior or interior surface per requirement of the system.
  - Synthesis of a catalyst
    - Chemicals necessary for catalyst preparation: cobalt nitrate hexahydrate and ammonium molybdate tetrahydrate are used as Co precursor and Mo precursor, respectively. Gamma-Al_2O_3 with high surface area (100×10^{-3} $m^2.Kg^{-1}$) and low surface area (60×10^{-3} $m^2.Kg^{-1}$) is used as catalyst support. Millipore water is used for slurry preparation.
  - Preparation of a catalyst:
    - Mix precursors with millipore water by maintaining a 1:1 water/solid ratio to get a slurry solution.
    - Keep at the ambient temperature (28°C ± 2°C) for 30 min to retain the homogeneity of the sample.
    - Heat at 120°C for 9 h to remove excess water from the solution.
    - Calcinate at 400°C for 5 h at a heating rate of 5°C/min.
  - Characterization techniques:
    - BET surface area analysis—to know pore size and distribution of pores
    - Particle size analysis—to measure particle size and its distribution
    - FTIR—to confirm formation of complex between precursors of Co, Mo, and alumina
    - XRD—to understand the change in phase of catalysts with the change in calcination temperature and dispersion of active phase

- – FESEM—to recognize the surface texture of the catalyst
- – ESR—to characterize the electronic structures of materials with unpaired electrons
- – TPR—to find the most efficient reduction conditions
- – CO chemisorption—to determine the metal dispersion, metallic surface area, and active particle diameter
- – LRS—to know the local structure of the oxidic precursors
- – TEM—to identify the morphologic, compositional, and crystallographic information of the synthesized catalyst

- Coating of catalyst on support membrane:
  - Preparation of MoS_2 + catalyst solution
  - Coating over the exterior of the membrane using the "paint coating" technique
    - – Study of membrane surface morphology—average pore diameter of the optimized catalytic membrane is found as 1.76×10^{-6} m and porosity 35%
    - – Film thickness—in the range of 9×10^{-8} to 9.5×10^{-8} m
    - – Distribution of catalyst over the membrane surface is well appreciable

- System economics—probable cost of the optimized catalytic membrane evaluated as \$381/m^2

References

1. S. Bose, C. Das, Preparation, characterization, and activity of activated γ-alumina-supported molybdenum/cobalt catalyst for the removal of elemental sulfur. *Applied Catalysis A: General*, 512 (2016) 15–26.
2. Mott Metallurgical Corporation, Farmington, CT (2007). http://www.mottcorp .com
3. Ch. Papadopoulou, J. Vakros, H.K. Matralis, Ch. Kordulis, A. Lycourghiotis, On the relationship between the preparation method and the physicochemical and catalytic properties of the CoMo/γ-Al$_2$O$_3$ hydrodesulfurization catalysts. *Journal of Colloid and Interface Science*, 261 (2003) 146–153.
4. C. Agrafiotis, A. Tsetsekou, Deposition of meso-porous g-alumina coatings on ceramic honeycombs by sol-gel methods. *Journal of the European Ceramic Society*, 22 (2002) 423–434.

8

Study of Mass Transfer of CMRs

8.1 Introduction

In this chapter, the removal or recovery of elemental sulfur using the fabricated catalytic membrane reactor (CMR) as a function of catalytic activity and mass transport is studied. A batch experiment is carried out at room temperature to collect elemental sulfur in solid form, by reacting hydrogen sulfide and sulfur dioxide. The recovery of elemental sulfur is confirmed by means of the catalytic activity (conversion, yield, and selectivity) and turnover frequency of the catalyst. The deposition of elemental sulfur on the catalytic layer of the membrane is recognized by Fourier transform infrared spectroscopy and x-ray diffractometry study. The multireactant mass transfer behavior of the CMR is studied on the basis of reaction conditions, membrane properties such as reaction equilibrium constant (K_{eq}), and membrane area and reactor volume, with claims that the fabricated reactor behaves like an ideal CMR by means of the mass transport.

8.2 Gas Absorption and Gas–Solid Catalyzed Reaction through CMRs

To study the feasibility of mass transport inside the fabricated CMR, a model has been adopted that is very similar to one in the study for multicomponent mass transfer at the bulk–fluid membrane interface in a CMR [1]. A novel reactor is fabricated as a tubular configuration (illustrated in Figure 8.1).

A CMR must not simply combine a membrane separation unit with a chemical reactor; it must also integrally couple them in such a way that a synergy is created between the two units, potentially resulting in enhanced performance in terms of separation, selectivity, and yield. This is the novelty of using a membrane reactor instead of a traditional packed-bed reactor where both reaction and separation occur simultaneously by eliminating undesired products, such as water, CS_2, and COS, and maintaining thermodynamic

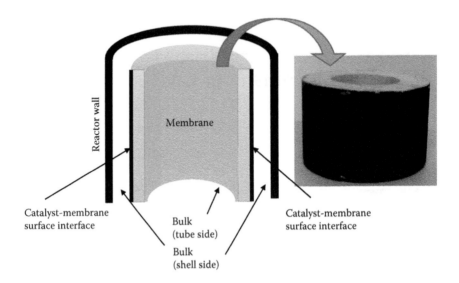

FIGURE 8.1
CMR configuration for multicomponent mass transport.

equilibrium. This CMR mainly consists of two compartments (e.g., shell-and-tube configuration). The H_2S and SO_2 gas (center of the reactor) are fed from the same direction at the shell side and tube side of the membrane simultaneously, as schematically represented in Figure 8.2. The H_2S gas is fed over the catalytic surface, whereas the SO_2 gas is diffused through the membrane pores from the tube side (no catalytic layer) to the shell side surface and reacts with H_2S over the catalytic layer. The product deposits over the exterior surface of the membrane. The purpose of feeding H_2S at the shell side is to maintain a minimum distance from the catalytic layer (reaction region) and to retain as much as with the reaction place, resulting in higher conversion and negligible slip of the reactant to the tube side.

During the operation, low pressure (0.19 MPa) is maintained at the shell side due to high diffusivity of H_2S (4.2×10^{-7} m^2.min^{-1}) and high pressure (0.39 MPa) is retained at the tube side owing to low diffusivity of SO_2 (1.34×10^{-7} m^2.min^{-1}). The entire process is performed at room temperature in batch mode. To the best of the authors' knowledge, there is no report on a Claus reaction over an activated γ-alumina (high surface area) based Mo–Co catalyst at room temperature. But, catalytic activity of catalysts at room temperature for different purposes has been reported over several noble metal catalysts, such as Pt/TiO$_2$, Pt/Fe$_2$O$_3$, etc. [2–4]. Furthermore, the fabricated catalytic membrane (thickness: 10×10^{-3} m; diameter: 50×10^{-3} m) of 45×10^{-3} m height is considered with a membrane support (pore diameter range ~ 0.3×10^{-6} to 0.8×10^{-6} m) with a microporous catalytic layer (pore diameter range ~ 20×10^{-9} to 84×10^{-9} m) on the top surface of the membrane.

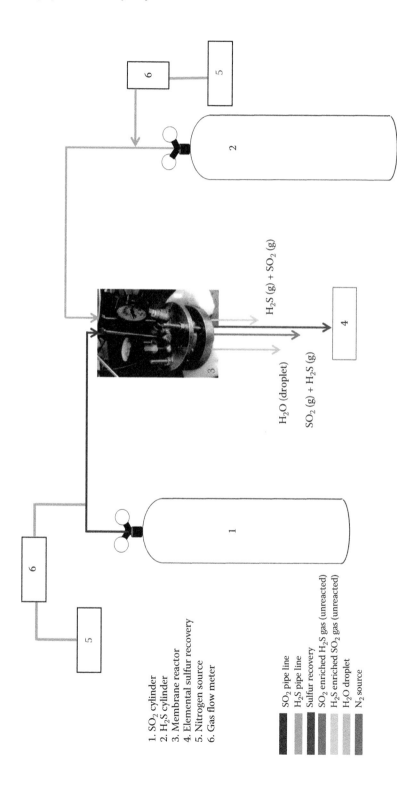

1. SO$_2$ cylinder
2. H$_2$S cylinder
3. Membrane reactor
4. Elemental sulfur recovery
5. Nitrogen source
6. Gas flow meter

SO$_2$ pipe line
H$_2$S pipe line
Sulfur recovery
SO$_2$ enriched H$_2$S gas (unreacted)
H$_2$S enriched SO$_2$ gas (unreacted)
H$_2$O droplet
N$_2$ source

H$_2$S (g) + SO$_2$ (g)

H$_2$O (droplet)

SO$_2$ (g) + H$_2$S (g)

FIGURE 8.2
Schematic diagram of the experimental setup.

8.3 Practical Example

8.3.1 Conversion of H₂S into Elemental Sulfur

The catalytic activity of the prepared and optimized catalysts by means of high metallic dispersion (MD) and metallic surface area (MSA) is measured by the Claus reaction. The catalytic experiments were carried out at atmospheric pressure at room temperature, using a batch-mode tubular reactor (inside diameter: 0.09 m; length: 0.05 m; volume: 2.86×10^{-4} m³), which consists of a catalytic membrane (outside is coated with the catalyst) situated inside the reactor. The feed consisted of a mixture of H_2S, SO_2, and N_2 with a volume percent of H_2S and SO_2 varying from 0.3 to 1.9 and 0.2 to 1. The initial concentrations of H_2S and SO_2 were 1.35×10^{-3} and 1.01×10^{-3} g.lit⁻¹, respectively. In the reaction procedure, both H_2S and SO_2 were fed into the reactor at a constant pressure of 3.45×10^{-3} MPa and reacted on the surface of the membrane, where exactly 4.6 g catalyst was coated and elemental sulfur deposited on the membrane surface. Pressure inside the membrane was maintained at 6.8×10^{-2} MPa. N_2 gas was used to keep the reactor environment inert. The permeate gas flux was taken from the bottom of the reactor at different time intervals. The catalytic activity of the prepared catalyst was measured by means of conversion of reactant, yield, and selectivity of the product and the turnover frequency (TF) of the catalyst. Mathematical expressions of the parameters are as follow:

$$\text{Overall conversion (\%)} = 100 \times \frac{C_{R_{inlet}} - C_{R_{outlet}}}{C_{R_{inlet}}} \tag{8.1}$$

$$\text{Yield (\%)} = 100 \times \frac{\text{The amount of sulfur produced}}{\text{Total amount of reactant fed}} \tag{8.2}$$

$$\text{Selectivity (\%)} = 100 \times (\text{Overall conversion} \times \text{Yield}) \tag{8.3}$$

$$\text{Turnover Frequency (TF)} = \frac{\text{Amount of product (mol)}}{\text{Amount of the catalyst active sites} \times \text{time}} \tag{8.4}$$

$$\text{Effective diffusitivity } (D_{ij}) = \frac{1.0133 \times 10^{-7} T^{1.75}}{P\left[\left(\sum v_A\right)^{1/3} + \left(\sum v_B\right)^{1/3}\right]^2} \left[\frac{1}{M_A} + \frac{1}{M_B}\right]^{0.5} \tag{8.5}$$

where

$C_{R_{inlet}}$ is the initial concentration of reactant gases in the inlet of the reactor (mol.m^{-3})

$C_{R_{outlet}}$ is the concentration of feed gases (mol.m^{-3}) present in the permeate

D_{ij} is the effective diffusivity (82.8 m^2.min^{-1})

T is the temperature (°C/K)

P is the total pressure (MPa)

M_A and M_B are the molar mass of H$_2$S and SO$_2$, 34.08 and 64.06 g.mol^{-1}, respectively

v_A and v_B are the volume terms of the reactant gases in m^3, 21.62 and 27.96, respectively

Individual conversion of the reactant gas was measured using Equation 8.1 by considering individual concentration of the reactant gas.

It was observed that the catalyst with high MD and MSA values showed a gradual increase in conversion, yield, and selectivity, whereas the catalyst having minimum MD and MSA showed an opposite trend (see Figure 8.3). In the case of a catalyst showing minimum MD and MSA, the conversion of hydrogen sulfide and sulfur dioxide was maximum at start of the reaction but decreased gradually with time (Figure 8.3b) and reached a limiting value opposite to the trend shown in Figure 8.3(a). This phenomenon clearly indicates a beginning of catalyst deactivation as the number of active metal sites is less in the catalyst having minimum MD and MSA. It may be stated here that the MD and MSA are greatly influenced by the preparation process and metal-support interaction and are responsible for the formation of the desired product by means of higher catalytic activity. It was also further observed that yield and selectivity of the elemental sulfur increased with the increase in overall conversion of the reactant gases at a decreasing rate of disappearance of reactant gases with time (Figure 8.3c); the maximum yield was 87%. Elemental sulfur was produced mainly from hydrogen sulfide species adsorbed on the metal (Mo) active sites, which were responsible for the catalytic reaction.

8.3.1.1 Another Example

Case study: The modified Claus process is the most common method for the conversion of hydrogen sulfide to sulfur contained in sour oil and natural gas. An important but relatively untouched part of the Claus process—the reaction kinetics of the front-end reaction furnace in which sulfur production takes place, hydrocarbon contaminants are destroyed, and reactions occur that prepare the sour gas for downstream catalytic processing—has been studied [5]. The kinetics of the second part of the Claus reaction, the reaction between H$_2$S and SO$_2$, has also been studied at Claus reaction furnace conditions, and new, consistent experimental data presented for H$_2$S and SO$_2$ conversions. All the experiments were carried out in a laboratory

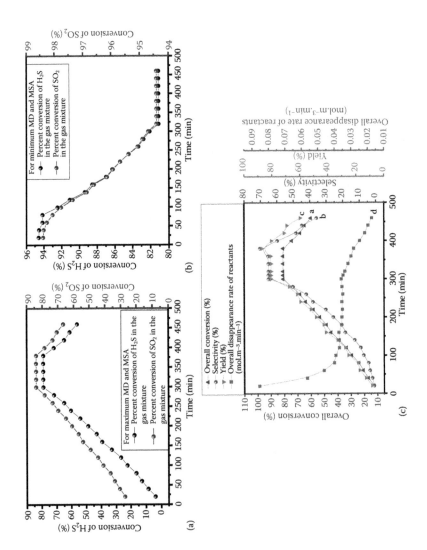

FIGURE 8.3
Conversion of H_2S into elemental sulfur as a function of time with the optimized catalyst: (a) having maximum MD (93.69%) and MSA (480.51×10^{-3} m^2 Kg^{-1}), (b) having minimum MD (1.15%) and MSA (5.94×10^{-3} m^2 Kg^{-1}), (c) relation between product concentration and overall conversion of reactant with time.

scale under isothermal conditions using a plug-flow reactor at temperatures between 850°C and 1150°C and at residence times between 0.05 and 1.2 s. In this experiment, H_2S and SO_2 were fed to the reactor and mixed at furnace temperatures. The following experimental conditions were maintained: inlet H_2S concentrations: 0.25% to 3%; inlet SO_2 concentrations: 0.3% to 4%; residence times: 50 to 1200 ms: and temperature: 850°C to 1150°C. The material balance based on measuring both H_2S and SO_2 and adjusting for H_2S cracking/reassociation showed the data to be consistent within an average of 25% over a residence time range from 0.1 to 0.9 s. Results showed both H_2S and SO_2 conversion as a function of residence time and temperature. It was observed that at 850°C, H_2S conversion was low, ranging from about 1%–5% for residence times ranging from 0.15 to 1.25 s. On the other hand, at 1150°C, H_2S conversion was relatively high, ranging from about 40% to 85% over the same residence time range. Similarly, the obtained SO_2 conversion was also low at 850°C, ranging from about 1% to 5% for residence times ranging from 0.25 to 1.1 s. Conversely, at 1150°C, SO_2 conversion was again moderately high, ranging from about 25% to 70% at residence times ranging from 0.1 to 0.63 s.

A new reaction rate expression for the second part of the Claus reaction from the new experimental data was proposed as follows:

$$r = A_f e^{E_{af}/RT} P_{H_2S} P_{SO_2}^{0.5} - A_r e^{E_{ar}/RT} P_{H_2O} P_{S_2}^{0.75}$$

where
  A_f is 15,700 ($\pm$1200) mol cm^{-3} s^{-1} atm$^{-1.5}$
  E_{af} is 49.9 ($\pm$0.3) kcal mol^{-1}
  A_r is 500 ($\pm$50) mol cm^{-3} s^{-1} atm$^{-1.75}$
  E_{ar} is 44.9 ($\pm$0.5) kcal mol^{-1}

8.4 Kinetics of Catalyst on Membrane

8.4.1 External Diffusion Resistance

In many industrial reactions, the overall rate of the reaction is limited by the rate of mass transfer of reactants between the bulk fluid and the catalytic surface. In this study, the external surface of the catalytic layer was surrounded by the reactant H_2S in the bulk and reacted with other reactant SO_2 diffusing through the membrane from the tube side; the reaction took place only on the external catalyst surface and not in the bulk surrounding it. To interpret the external diffusion resistance, a model equation was utilized and will be discussed later.

8.4.2 Internal Diffusion Resistance

Internal diffusion resistance was determined to examine the effects of diffusion of the reactants into the pores within the catalyst pellet on the overall rate of reaction. In the case of a porous catalyst, diffusional limitations can change the overall kinetics of a reaction through internal and external resistances. Established criteria and experimental methods can determine the presence or absence of internal mass transfer limitations. The objective here was to apply a realistically accurate analysis based on observed variable properties of the reactant and physical properties of the catalyst to determine the presence or absence of any significant internal mass transfer limitations. The most suitable approach was the Weisz–Prater criterion (φ_{W-P}), as the reaction followed pseudo first-order kinetics [6], which represents a ratio of the rate of the reaction to the rate of diffusion in the pores. This criterion was calculated using the following equation and the value was obtained as 0.0019, which is less than 0.6:

$$\text{Weisz-Prater criterion } (\varphi_{W-P}) = \frac{r_a R_p^2}{C_s D_{ij}} \tag{8.6}$$

where
  r_a is the reaction rate per volume of the catalyst (1.51×10^{-4} mol.m^{-3}.min^{-1})
  R_p is catalyst particle radius (0.133×10^{-6} m)
  C_s is the reactant concentration at the external surface of the catalyst (1.68×10^{-3} mol.m^{-3})

The reason behind yielding a smaller φ_{W-P} value is the equimolar consumption of H_2S and SO_2 by the reaction, as there is no significant secondary reaction to form intermediate products like carbonyl sulfide and carbonyl disulfide within the first sampling time. Therefore, internal pore diffusion limitation is not responsible for the reduction in conversion of H_2S after 380 min of reaction and can be neglected for the kinetic study of the Claus reaction.

8.5 Reaction Kinetics and Mass Balance Equation

The Claus reaction on the catalytic membrane is described with the help of mathematical equations in which the mass transfer from the gas bulk to the catalyst–membrane surface is calculated according to the Maxwell–Stefan model. The Claus reaction is considered as the oxidation of H_2S by SO_2 over the catalytic surface of the membrane, which forms elemental sulfur by the reaction pathway given in Table 8.1 [7].

TABLE 8.1

Sulfur Formation Steps for the Oxidation of H_2S by SO_2 (Claus Reaction)

$H_2S + Z \leftrightarrow Z\text{–}SH_2$	(a)
$SO_2 + Z \leftrightarrow Z\text{–}SO_2$	(b)
$Z\text{–}SH_2 + Z \rightarrow Z\text{–}SH + Z\text{–}H$	(c)
$Z\text{–}SO_2 + Z\text{–}H \rightarrow Z\text{–}S: + Z\text{–}O_2H$	(d)
$Z\text{–}SH + Z\text{–}O_2H \rightarrow Z\text{–}S: + Z\text{–}HO_2H$	(e)
$Z\text{–}HO_2H + Z \rightarrow 2Z\text{–}OH$	(f)
$Z\text{–}SH + Z\text{–}OH \rightarrow Z\text{–}S: + Z\text{–}OH_2$	(g)
$Z\text{–}OH + Z\text{–}H \rightarrow Z\text{–}OH_2 + Z$	(h)
$Z\text{–}OH_2 \leftrightarrow Z + H_2O$	(i)
$2Z\text{–}S: \leftrightarrow Z\text{–}SS: + Z$	(j)
$2Z\text{–}SS: \leftrightarrow Z\text{–}SSSS: + Z$	(k)
$2Z\text{–}SSSS: \leftrightarrow Z\text{–}SSSSSSSS: + Z$	(l)
$Z\text{–}SSSSSSSS: \leftrightarrow Z + S_8$	(m)
Z – active site on the catalyst surface	(n)

Source: R. Taylor, R. Krishna, *Multicomponent mass transfer*, Wiley series in chemical engineering, John Wiley & Sons, New York, 1993.

The homogeneous Claus reaction is as follows:

$$2H_2S + SO_2 \underset{k_b}{\overset{k_f}{\rightleftharpoons}} \frac{3}{x} S_x + 2H_2O \tag{8.7}$$

where x is 2 or 6 or 8 and k_f and k_b are the forward and backward rate constants.

As the entire reaction occurs in the presence of a heterogeneous catalyst, the scheme has to be considered as a heterogeneous process. But process calculations such as reaction kinetics are measured considering the system as a homogeneous system due to the consideration of mixed flow patterns of the reacting fluids [8]. In an ideal contacting of a heterogeneous system, each fluid is retained in the reactor as mixed flow. In this study, as one of the phases was discontinuous, as a droplet, its macro fluid characteristics had to be accounted for because the material was in mixed flow. To avoid difficulties during calculation, we have neglected macrofluid characteristics of the product and calculated all the parameters of the system considering a homogeneous system.

A material balance for the batch reactor was executed to verify the change in rate of disappearance of reactant gases with time. It was assumed that no fluid entered or left the reaction mixture during reaction as it was operated in batch mode. Therefore, the material balance in the reactor was as follows:

$$-r_{overall} = C_{overall_0} \cdot \frac{dX_{overall}}{dt} \tag{8.8}$$

where
  t is the time required to convert reactant gases into products (min)
  $C_{overall_0}$ is the overall initial concentration of both the reactants (mol.m^{-3})
  $-r_{overall}$ is the rate of disappearance of the reactant gases (mol.m^{-3}.min^{-1})
  $dX_{overall}$ is the change in conversion of the reactants into product with time

The permeation rate via the membrane was calculated according to the dusty gas model [9] as the driving force was the pressure difference across the membrane. For simplicity of the mathematical model, the outer surface of the membrane was considered to be the effective area of the membrane where both the reaction and separation occur simultaneously; the effective area of the membrane was thus determined by the height and the outer radii of the membrane:

$$\text{Effective membrane area } (A_{eff}) = 2\pi r_o h \tag{8.9}$$

where
  A_{eff} is the effective membrane area where both the reaction and separation
    occur in m^2
  r_o is the outer radius of the membrane coated with catalyst in m
  h is the height of the membrane in m

The mass balance for each component across the membrane–catalyst interface (shell side) is given by [1]

$$(v_\phi x_i c_t)_{in} - (v_\phi x_i c_t)_{out} + v_i VR_b - A_{eff} J_i = 0 \tag{8.10}$$

where
  v_ϕ is the volumetric flow rate (m^3.min^{-1})
  x_i is the molar fraction of the species and i the interface
  c_t is the total concentration of the reactant gases (mol.m^{-3})
  v_i is the stoichiometric coefficient of component i
  V is the reactor volume (m^3)
  R_b is the rate of the reaction in the bulk (mol.m^{-3}.min^{-1})
  J_i is the flux of ith component gases (mol.m^{-2}.min^{-1})

As the reactor was operated in batch mode, no fluid was entering or leaving during the reaction. Therefore, Equation 8.10 becomes

$$v_i VR_b - A_{eff} J_i = 0 \tag{8.11}$$

In the CMR, the reaction in the bulk was neglected as the reaction predominantly took place at the catalytic layer on the membrane at the catalyst–membrane surface interface. Therefore, the relation for the reaction rate at the interface of the membrane is given by [1]

$$R_{int} = k_f \left(\overset{reactants}{\prod_j} (c_t x_i)^{|v_j|} - \frac{1}{K_{eq}} \overset{products}{\prod_j} (c_t x_i)^{v_j} \right) \tag{8.12}$$

Here, R_{int} is the rate of the reaction at the interface (mol.m^{-3}.min^{-1}); k_f is the forward reaction rate constant (mol. L^{-1}.min^{-1}); K_{eq} is the equilibrium constant of the reaction, which is assumed to be valid through the entire membrane; and v_j is the stoichiometric coefficient of component j.

The difference in mole fraction of H$_2$S in the bulk and the catalyst–membrane surface interface through the boundary layer was calculated using the Maxwell–Stefan equation [10]. The Maxwell–Stefan theory was also applied to verify the diffusion of reactant gases (H$_2$S and SO$_2$) along with undesired product (H$_2$O) through the membrane. The importance of this equation is to verify any diffusion of water (undesired product), sulfur, and unconverted reactant gas on the membrane wall during reaction by observing the change in mole fraction of H$_2$S:

$$-c_t \Delta x_i = \sum_{j \neq i} \frac{\overline{x_j} J_i - \overline{x_i} J_j}{k_{ij}} \tag{8.13}$$

where
   Δx_i is the difference in mole fraction between the bulk and the catalyst–membrane surface interface
   k_{ij} is the mass transfer coefficient for components i and j (m.min^{-1})
   $\overline{x}$ is the average mole fraction between the bulk and the catalyst–membrane surface interface

Assume the summation of molar fractions of SO$_2$ at the bulk (x_j) and the interface $\left(x_j^{int} \right)$ is equal and is given by [1]

$$\sum_{j=1}^{nc} x_j = \sum_{j=1}^{nc} x_j^{int} = 1 \tag{8.14}$$

Additional equations are provided to calculate K_{eq} and k_{ij}:

$$K_{eq} = \frac{x_{S_8}^{\frac{3}{8}} x_{H_2O}^2}{x_{H_2S}^2 x_{SO_2}} p^{\frac{-5}{8}} \tag{8.15}$$

where x_{S_8}, x_{H_2O}, x_{H_2S} and x_{SO_2} are the mole fractions of products and reactants, respectively:

$$k_{ij} = \frac{D_{ij}}{l_d} \qquad (8.16)$$

where
l_d is the distance between the reactor wall and the membrane surface where the diffusion occurs (in m)
p is pressure at which reactant gases enter the reactor (in MPa)
D_{ij} is the effective diffusivity of the reactant gases (m².min⁻¹), determined through the Füller–Schettler–Giddings equation [11] (i.e., Equation 8.5)

Before solving Equations 8.11 through 8.16 simultaneously, the overall flux has to be determined. In this study, it was assumed that there are two mechanisms of transport: viscous flow and molecular diffusion according to Fick's law. Thus, flux through the boundary (catalytic) layer of the membrane was calculated by adding the flux of H₂S through the membrane wall (catalyst–membrane interface) and the viscous flux due to the flow of compressible fluid through a porous support membrane, considering that they are independent:

$$J = J_{membrane,i} + J_{visc} \qquad (8.17)$$

The flux through the membrane is calculated from

$$J_{membrane,i} = P_i x_i P_{tot} \qquad (8.18)$$

where
P_i is the permeance (mol.m⁻².min⁻¹.MPa)
x_i is the mole fraction of the species i at the interface
P_{tot} is the total pressure (MPa) for this multicomponent system

It was assumed that the partial pressure of the permeating components at the permeate side was negligible. This is close to reality because a vacuum or a sweep gas is often used at the permeate side, which removes all permeating species and keeps the partial pressure at the permeate side at the very low range. Pressure through the tube side to the shell side of the membrane was the driving force for this study. Therefore, depending on the rate of diffusion of reactant gases, pressure was chosen as one of the significant factors for membrane flux calculation. As SO₂ (in the tube side) was less diffusive than H₂S, more pressure was required to diffuse SO₂ through the tube side to the shell side. Otherwise, after feeding, due to a high rate of diffusion of H_2S, the gas would immediately diffuse through the membrane wall from the shell side to the tube side and prevent product formation. Concentration in terms of mole fraction was considered another important factor due to the dependency of membrane permeate flux on reaction equilibrium constant. It was assumed that viscous flux includes both Knudsen diffusion and viscous flow due to the distribution of wide ranges of pores and a strong affinity of the membrane surface for the reactant gas molecules to be transported through the core of the support membrane to the catalytic layer and vice versa:

$$J_{visc} = -\frac{1}{RT}\left(D_{inert,k} + \frac{B_0 p}{\eta}\right)\frac{dp}{dx} \tag{8.19}$$

where

$J_{membrane,i}$ is the flux of the reactant gases through the membrane wall (mol.m^{-2}.min^{-1})

J_{visc} is the flux considering transport mechanism, viscous flow (mol.m^{-2}.min^{-1})

R is the universal gas constant (m^3.MPa.K^{-1}.mol^{-1})

$D_{inert,k}$ is the Knudsen diffusion coefficient (m^2.min^{-1})

η is the viscosity of the gas (MPa.min)

t is the time (min)

The viscous flow is represented by Equation 8.19 for the flow of a compressible gas through a porous body and is solved by integrating over the thickness of the membrane ($L = 0$ to 0.001 m) and the reaction time (0 to 460 min). The parameter B_0 is a specific membrane parameter (m^2), defined by

$$B_0 = \frac{\varepsilon}{\tau}\frac{d_p^2}{32} \tag{8.20}$$

$$D_{inert,k} = \frac{4}{3}\frac{d_p}{3}\sqrt{\frac{4RT}{\pi M_{N_2}}} \tag{8.21}$$

where

ε is the porosity

τ is the tortuosity factor

d_p is the pore diameter (m)

M_{N_2} is the molecular weight of the inert gas (kg.mol^{-1})

8.6 Influence of Reaction Rate, Equilibrium Reaction Constant, and Mass Transport Coefficient at Membrane Boundary Layer

The conversion of H_2S into elemental sulfur depends on time, equilibrium reaction constant, mass transport coefficient, and mostly the removal of water. Figure 8.4 describes control of the conversion as a function of equilibrium constant and reaction rate as disappearance of H_2S. The variation in equilibrium constant K_{eq} (from low to intermediate) over different ranges of time indicates formation of sulfur on the membrane surface. The asymptotic nature of the equilibrium constant value (curve a) has three distinct sections with reaction time, labeled PQ, QR, and RS. In section PQ, a gradual increase of K_{eq} up to 320 min is observed due to high rate of reaction. The molecules of reactant gases adsorb onto the catalytic surface, attach to the active sites and react, indicating a reaction-controlled zone. In section QR, the equilibrium constant value remains constant from 320 to 400 min because mass transfer effects are not important as reaction rate is limiting. The section RS represents a sharp equilibrium falling zone, which specifies fast rate of mass transfer compared to the reaction rate, corresponding to a pseudo-irreversible reaction.

These results suggest that conversion is not affected by the undesired product—that is, the production of water but a significant effect of K_{eq} and k_{ij}. Though the performance of the CMR reduces after a certain period of

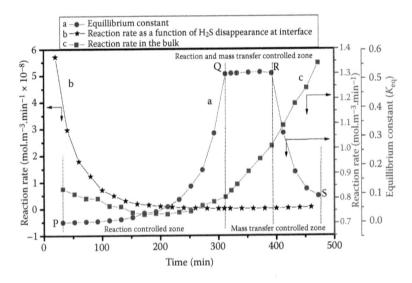

FIGURE 8.4

Variation of equilibrium constant with time (curve a), reaction rate as a function of disappearance of H_2S evaluated using Equation 8.6 (curve b), and overall reaction rate calculated on the basis of mass balance using Equation 8.7 (curve c).

time, a higher conversion is achieved due to the shift of equilibrium from reactant to the product side for low and intermediate K_{eq} value. As reaction time increases, reactant particles come closer to each other and convert the reactants into the desired product.

Figure 8.4 (curves b and c) shows the variation in reaction rate in the bulk and at interface with time by means of mass balance equation and forward reaction rate constant K_f as a function of the equilibrium constant, respectively. Figure 8.4 (curve b) indicates a decrease in reaction rate as a function of disappearance of H_2S with time at the interface of the membrane, suggesting occurrence of reaction at the catalytic surface of the membrane. The declining nature of the reaction rate can also be described by the increase in viscous flux, as is seen in Figure 8.4 (curve a). Figure 8.4 (curve c) shows an increase in reaction rate with time and K_f, but the reaction kinetics is slow as K_f lies in the range of 10^{-2} m.min^{-1}, causing a small amount of water production. The nature of this increase resembles the increase in the equilibrium constant, which is in the favor of the CMR.

The relationship between output factors (overall conversion, selectivity, and yield) with time obtained from the experimental data was compared with the predicted values on the basis of a polynomial fitting. A statistical analysis of variance (ANOVA) was performed to validate the experimental values with the theoretical values.

The predicted values obtained from the polynomial fitting (Tables 8.2 and 8.3) were compared with the values obtained from experimental studies

TABLE 8.2

Polynomial Fitting of Catalytic Activity (Experimental Data) of the CMR with Theoretical Data

Description	Perform Polynomial Fitting
Equation	$Y = intercept + B1 \times X^1 + B2 \times X^2$
Weight	No weighting
Multidata fit mode	Independent fit—consolidated report

TABLE 8.3

Determination of Catalytic Activity of the CMR on the Basis of Polynomial Fitting

		Value	Standard Error
Overall conversion	Intercept	9.97	1.15
	B1	0.19	0.02
	B2	1.34×10^{-4}	4.34×10^{-5}
Selectivity	Intercept	1.79	1.50
	B1	−0.02	0.02
	B2	7.11×10^{-4}	5.66×10^{-5}
Yield	Intercept	7.36	1.55
	B1	0.15	0.02
	B2	2.75×10^{-4}	5.89×10^{-5}

(actual values) to confirm the stability and the feasibility of the empirical model (Figure 8.5). Figure 8.5(a) depicts that the proposed model is well accepted and follows a polynomial fitting with an adjusted R-squared value of 0.99 (Table 8.4) for overall conversion, selectivity, and yield. The standard error for the output parameters lies within the range of ±6% (Table 8.5). Figure 8.5(b) presents the plot of regular residuals for overall conversion, selectivity, and yield, which reveals a random scattering of all experimental data points across the horizontal line or residuals. This phenomenon suggests that the proposed model is adequate and is also confirmed by the ANOVA results, as shown in Table 8.6.

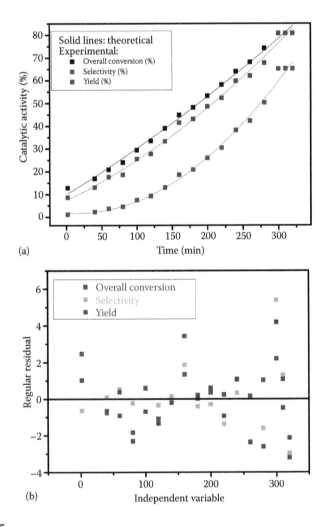

(a)

(b)

FIGURE 8.5
(a) Plot of experimental and theoretical values for overall conversion, selectivity and yield, and (b) plot of regular residual for overall conversion, selectivity, and yield.

TABLE 8.4

Statistical Analysis of the Predicted Catalytic Activity of the CMR

	Overall Conversion	Selectivity	Yield
Number of points	17	17	17
Degrees of freedom	14	14	14
Residual sum of squares	32.10	54.47	58.88
Adj. R-square	0.99	0.99	0.99

The new experimental data from the preceding study by means of a newly proposed catalyst synthesis scheme and H_2S conversion using Claus reaction were compared with the values obtained from the conventional methods reported in previously published literature [5,12–14]. A CoMo/γ-Al_2O_3 catalyst was prepared by equilibrium deposition filtration (EDF) and dry impregnation and compared with three CoMo/γ-Al_2O_3 samples prepared using various conventional impregnation methods. The authors found 30%–43% more conversion in hydrodesulfurization using the EDF-prepared catalyst than those prepared with the conventional impregnation techniques—almost 60% lower than the experimental data achieved in this study [13]. Conversions of H_2S and SO_2 at a temperature of 276°C and a pressure of 1.5 bar in absence of a pressure difference over the membrane using an α-alumina-based membrane of 41% porosity and 45 mm of thickness were measured. The aim was to present the basic fundamentals of the membrane reactor—not to present the parameter values [12]. Almost 63% and 82% H_2S conversion with a different rate expression and reaction conditions at 850°C and 1150°C, respectively, was found. In a similar study with a different rate expression but at the same reaction temperatures of 850°C and 1150°C, 12% and 85% H_2S conversion was observed—lower than the experimental data obtained in this study [5].

In this study, SO_2 conversion was also measured and these values followed almost similar trends with H_2S conversion. The obtained data were consistent and suggest that the differences in results between this study and previous studies are due to a new proposed scheme and negligible mass transfer limitation.

8.7 Summary

An attempt to fabricate a catalytic membrane and membrane reactor for the removal or recovery of elemental sulfur using the Claus reaction was successfully completed and vital conclusions drawn in four major segments of the entire work: (a) fabrication, characterization, and optimization study of

TABLE 8.5

Summary of Predicted Catalytic Activity Based on Statistical Analysis

	Intercept		B1		B2		Statistics	
	Value	Standard Error	Value	Standard Error	Value	Standard Error	Value	Standard Error
Overall conversion	9.97	1.15	0.19	0.02	1.34×10^{-4}	4.35×10^{-5}	0.99	9.96
Selectivity	1.78	1.50	-0.020	0.02	7.11×10^{-4}	5.67×10^{-5}	0.99	1.78
Yield	7.36	1.55	0.15	0.022	2.75×10^{-4}	5.89×10^{-5}	0.99	7.35

TABLE 8.6

ANOVA for Overall Conversion, Selectivity, and Yield

		DF	Sum of Squares	Mean Square	F Value	Prob > F
Overall conversion	Model	2	8808.26	4404.13	1920.68	0
	Error	14	32.10	2.30		
	Total	16	8840.36			
Selectivity	Model	2	8556.26	4278.13	1099.59	4.44×10^{-16}
	Error	14	54.47	3.90		
	Total	16	8610.73			
Yield	Model	2	9472.45	4736.22	1126.19	3.33×10^{-16}
	Error	14	58.88	4.20		
	Total	16	9531.32			

a low-cost ceramic support membrane; (b) synthesis and characterization of the catalyst; (c) fabrication and characterization of a low-cost catalytic membrane; and (d) performance of catalytic membranes and membrane reactors.

The findings on the various issues examined in the fabrication, characterization, and application of catalytic membranes and membrane reactors are summarized as follows:

- For the commercial aspects, use of low-cost ceramic support for reducing the fabrication cost of catalytic membranes and membrane reactors has been considered as one of the key criteria. Incorporation of sawdust as a pore-former in the fabrication of low-cost ceramic support membranes, a scorching issue at the time, has been done efficiently. With respect to morphology, thermal and chemical stability, mechanical strength, and economic sustainability factors maintained for the ceramic support membrane, all the factors have been given equal emphasis. The fabricated tubular ceramic support costs of $332/m², including membrane mold preparation charges, are economical compared to the membranes available in the market and reported in the literature.

- A novel Mo–Co/activated γ-Al$_2$O$_3$ catalyst has been simply and effectively synthesized for the fabrication of catalytic membranes and membrane reactors. An excellent metal-support interaction has been observed in terms of well-distributed dispersion of metal precursors over a support matrix.

- Again, for the economic aspects, an inexpensive and innovative process (i.e., "paint coating" using solid lubricant over the fabricated low-cost ceramic support) has been considered as one of the key criteria for minimizing the manufacturing cost of catalytic membranes and membrane reactors. The fabrication of catalytic membranes is fruitful in context of morphology and budget. The estimated cost of the fabricated catalytic membranes of $381/m² proposed by the group of authors is still lower compared to costs for catalytic membranes reported in the literature.

- For environmental aspects, the aim of this study was extremely effective. The fabricated reactor behaved like an ideal CMR by means of the mass transport, where no internal pore diffusion mass transfer was observed during the reaction. A maximum of 87% yield and 80% conversion of the reactant into elemental sulfur were detected.

The results of this study confirm a productive fabrication of a best performing ceramic membrane reactor for environmental, commercial, and economic aspects. We believe that this work has large potential for future applications because it can minimize investment costs and intensify energy efficiency.

The discussion, so far, has focused on the fabrication and characterization of low-cost catalytic membrane reactors and their application. This study precisely established a process of minimizing manufacturing cost of a catalytic membrane that is novel and innovative.

This section delivers a number of future directions that would be valuable for further study of the fabrication and application of catalytic membrane reactors. Some essential areas of recommended research are suggested as an extension of the previously mentioned study:

- Reactor performance study can be carried out by changing different parameters, such as membrane thickness and membrane effective area.
- Modification of present batch setup (cocurrent) in other modes of operation, such as continuous countercurrent, can be done to obtain more realistic ideas upon the performance of the fabricated CMR in industrial applications.
- Modeling of reactors can be studied to get an accurate knowledge about reactor performance.

Despite the progress recognized in this area, further advances must expect the development of more stable and affordable catalytic membranes and membrane reactors. The probable benefits of such advances in the field of catalytic membrane and membrane reactors are the significant factors for possible large-scale applications in the oil and petrochemical industries. For this reason, the authors have carefully designed and analyzed catalytic membranes and membrane reactors to evaluate their potential advantages over the conventional process currently in operation. On account of high operating and manufacturing costs, the proposed scheme of using catalytic membrane reactors for sulfur recovery offers low capital investment and operating, manufacturing, and maintenance costs.

References

1. H. Mengers, N.E. Benes, K. Nijmeijer, Multi-component mass transfer behavior in catalytic membrane reactors. *Chemical Engineering Science*, 117 (2014) 45–54.
2. H. Huang, D.Y.C. Leung, D. Ye, Effect of reduction treatment on structural properties of TiO_2 supported Pt nanoparticles and their catalytic activity for formaldehyde oxidation. *Journal of Materials Chemistry*, 21 (2011) 9647–9652.
3. N. An, Q. Yu, G. Liu, S. Li, M. Jia, W. Zhang, Complete oxidation of formaldehyde at ambient temperature over supported Pt/Fe_2O_3 catalysts prepared by colloid-deposition method. *Journal of Hazardous Materials*, 186 (2–3) (2011) 1392–1397.
4. B.-B. Chen, C. Shi, M. Crocker, Y. Wang, M-.A. Zhu, Catalytic removal of formaldehyde at room temperature over supported gold catalyst. *Applied Catalysis B: Environmental*, 132–133 (2013) 245–255.
5. W.D. Monnery, K.A. Hawboldt, A. Pollock, W.Y. Svrcek, New experimental data and kinetic rate expression for the Claus reaction. *Chemical Engineering Science*, 55 (21) (2000) 5141–5148.
6. S. Mukherjee, M.A. Vannice, Solvent effects in liquid-phase reactions: I. Activity and selectivity during citral hydrogenation on Pt/SiO_2 and evaluation of mass transfer effects. *Journal of Catalysis*, 243 (1) (2006) 108–130.
7. R. Taylor, R. Krishna, *Multicomponent mass transfer*, Wiley series in chemical engineering, John Wiley & Sons, New York, 1993.
8. K. Sendt, B.S. Haynes, Role of direct reaction H_2S+SO_2 in the homogeneous Claus reaction. *Journal of Physical Chemistry A*, 109 (2005) 8180–8186.
9. E.A. Mason, A.P. Malinauskas, *Gas transport in porous media: The dusty gas model*, Elsevier, Amsterdam, the Netherlands, 1983.
10. J.A. Wesselingh, R. Krishna, *Mass transfer in multicomponent mixtures*, VSSD, Delft, the Netherlands, 2000.
11. E.N. Fuller, P.D. Schettler, J.C. Giddings, New method for prediction of binary gas-phase diffusion coefficients. *Industrial Engineering Chemistry*, 58 (5) (1966) 18–27.
12. H.L. Sloot, G.F. Versteeg, W.P.M. Van Swaaij, A non-permselective membrane reactor for chemical processes normally requiring strict stoichiometric feed rates of reactants. *Chemical Engineering Science*, 45 (8) (1990) 2415–2421.
13. Ch. Papadopoulou, J. Vakros, H.K. Matralis, G.A. Voyiatzis, Ch. Kordulis, Preparation, characterization, and catalytic activity of $CoMo/gamma-Al_2O_3$ catalysts prepared by equilibrium deposition filtration and conventional impregnation techniques. *Journal of Colloid and Interface Science*, 274 (1) (2004) 159–166.
14. S. Bose, C. Das, Preparation, characterization, and activity of activated γ-alumina-supported molybdenum/cobalt catalyst for the removal of elemental sulfur. *Applied Catalysis A: General*, 512 (2016) 15–26.

9

Various Applications of Ceramic Membranes

9.1 Introduction

Various applications of ceramic membranes as membrane contactors are discussed in this chapter. It includes five different sections where specific examples of applications are provided and discussed. Section 9.2 reports on gaseous stream treatment, particularly removal of different air pollutants such as volatile organic compounds (VOCs), acid gases, and SO_2 and mercury. The liquid stream treatment dealing with conventional applications of ceramic membranes for wastewater treatment, fruit juice clarification, and heavy metal separation is discussed in Section 9.3, as well as membrane contactors for the recovery of aroma compounds. A novel application of ceramic membrane in a fuel cell is also furnished in Section 9.4. In Section 9.5, other applications, including new categories of use of ceramic membranes, are described. Finally, in Section 9.6, commercial applications are demonstrated.

9.2 Gaseous Stream Treatment

9.2.1 VOC Removal

Nowadays, VOCs are of major environmental concern in our atmosphere and several methods of controlling them exist. Most methods have direct or indirect environmental consequences in terms of the materials they use and wastes they produce, albeit they control VOCs. It has been found that orthodox techniques, particularly those of physical and chemical natures, pose a number of difficulties such as high cost, high material requirement, and lots of waste that causes changes in climate and environmental degradation.

In a recent study, a dual membrane contactor containing two types of membranes (polypropylene and silicone rubber) was designed and constructed for the gas absorption process (Figure 9.1). The module's performance was evaluated based on permeation flux experiments. The experimental results

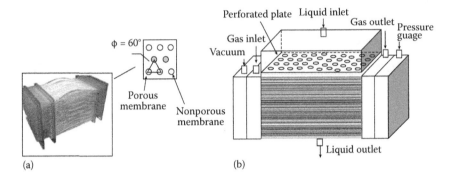

FIGURE 9.1
(a) Schematic drawing of membrane core and arrangement of dual hollow fiber membranes in lab scale module; (b) schematic drawing of dual membrane contactor.

were compared with the predictions from a numerical model. The mass transfer resistance in the fabricated module was also investigated using a resistance-in-series model. An empirical correlation was developed to describe the mass transfer coefficient under different operational conditions.

The performance tests of the proposed novel dual membrane design showed that by simultaneously stripping the gaseous component from the solvent through nonporous membrane along with the absorption process, permeation flux was improved by 12%–78%, compared to conventional modules containing only one type of porous membrane. It has been reported that the enhancement can be further improved by optimizing the membrane design and operational conditions. The effect of various operational conditions on absorption performance of the dual membrane module has also been examined. The experimental results were found to be in good agreement with those generated by simulation model. It was found that mass transfer resistance in the dual membrane module existed mainly in the liquid phase, while the membrane's resistance contribution to the total mass transfer resistance was about 7%. Higher permeation flux was achieved when the module was operated with lower gas flow rates, higher liquid flow rates, and a low vacuum pressure in nonporous membrane. When compared to dilute feed gas, removal efficiency with concentrated gas was much higher when other operational conditions are kept the same. This novel membrane design holds great potential for gas processing at remote locations such as offshore platforms [1].

In recent studies, various biological oxidation methods adopted in various reactors for controlling gaseous VOCs have been reviewed. Advantages and disadvantages of these technologies have also been studied and the delicate parameters for enhancing removal efficacy along with their environmental inferences recognized. It has been found that a rotating biological contactor (RBC) can be an effective substitute to control gaseous VOCs because of its simple design and operation. Also, it provides higher oxygen transfer, offers better mixing, consumes low energy, and has the potential of higher removal efficiency as compared to the other biological techniques. There is, moreover, an ample scope in improving its gaseous-control efficiency by optimizing parameters such as hydraulic loading rate, rotational speed, and hydraulic retention time, and similarly by use of different media, biofilm character-istics, and dissolved-oxygen levels. With these scopes, RBC can be an eco-friendly and promising choice for controlling gaseous VOCs [2].

Researchers have investigated the performance of a combined absorption–stripping process for removing dichloromethane, toluene, acetone, and methanol from air using an absorber, a polypropylene microporous hollow fiber, and a desorber containing hydrophobic polypropylene hollow fibers with an ultrathin and highly VOC-permeable plasma polymerized non-porous silicone skin on the outer surface. The absorber has been tested with silicon oil and Paratherm™ (adsorbents) and the desorber works by apply-ing a vacuum at the tube side. Excellent removal of VOC by the proposed system was reported. It has also been reported that the removal of species with higher Henry's value, such as dicholoromethane, is low for the coupled absorption–desorption system [3]. In another study, the performance of a hybrid system where the membrane-based absorption process was coupled to the membrane-based vapor permeation process was assessed. The hybrid system led to very high removal of methylene chloride (99.97%) [4].

In another recent study, highly hydrophobic titania, alumina, and zirconia porous ceramic membranes were prepared by grafting of $C_6F_{13}C_2H_4Si(OEt)_3$ (C6) and $C_{12}F_{25}C_2H_4Si(OEt)_3$ (C12) molecules and subsequently applied in a pervaporation (PV) process for removal of VOCs (methyl tert-butyl-ether/MTBE and ethyl acetate/EtAc) from binary aqueous solutions. All examined membranes were selective toward organic compounds. With an extended grafting time, a decrease of selectivity of the membranes was observed. The highest efficiency was found for the TiO_2 membranes independently of the utilized system and applied grafting molecules. Better transport and selec-tive properties were detected for membrane modified by C6 molecules than by C12 ones. The least effective membrane was alumina one. It has been observed that in the presence of organic compounds in the aqueous solution, the conformation of hydrophobic chains changes from tangled to straight [5].

9.2.2 Acid Gas Removal

Acid gases, especially, carbon dioxide (CO_2), are the major greenhouse gases contributing to global climate change. They are regularly present in gaseous streams and their exclusion by membrane contactors has been extensively investigated.

Research has paid considerable attention to membrane contactors since the 1980s, with Qi and Cussler the first to use hollow fiber membrane contactors for CO_2 absorption [6,7]. In addition to that, membrane contactors can capture other gases/vapors, such as SO_2 [8], and H_2S [9]; they have also been effectively commercialized in a number of industries, such as the semiconductor (production of ultrapure water), wine (CO_2 and O_2 removal), and membrane distillation industries [10].

Membrane contactors have significant advantages over conventional absorption systems. Those advantages include, among other things, better yields/selectivities (e.g., via equilibrium shift), better energy management, more compact design, extension of catalyst lifetime, etc. [10,11–13]. Those functions are schematically shown in Figure 9.2. These are nondispersive contacting arrangements in which the membranes do not offer selectivity for separation; however, as an alternative, these act as barriers to separate two phases and intensify the effective contact area for mass transfer. One of the most visible benefits of membrane contactors is their extremely high interfacial area, which can considerably cut equipment size and hence lead

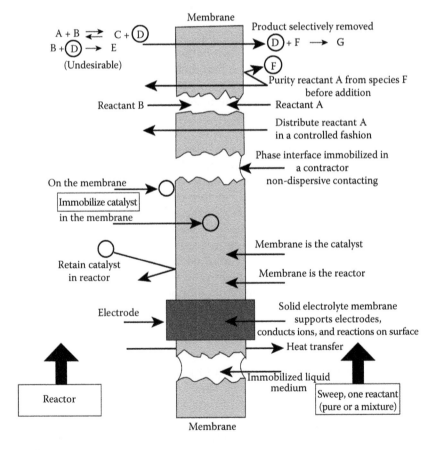

FIGURE 9.2
Membrane functions in chemical reactor systems.

to process intensification [7,11,14–16]. For example, the use of gas absorption membrane (GAM) systems has been explored as an alternative to traditional packed columns [16]. A membrane gas/liquid contactor has been used for natural gas sweetening (CO_2 and H_2S removal), dehydration, and CO_2 removal from exhaust gas. Membrane gas/liquid contactors operate with liquid on one side of a membrane and gas on the other. Here the pressure is essentially the same on both sides of the membrane and absorption into the liquid provides the driving force. The increased specific area of the membrane gas/liquid contactor allows for a 65%–75% reduction in weight and size compared to conventional towers [14].

A good example of the advantages of a membrane contactor is a CO_2 removal membrane contactor unit for the absorption of exhaust gas from an LM2500 gas turbine on an offshore installation reported in the literature [15]. The use of membrane contactors in this process eliminates or significantly reduces many operating problems, such as foaming and corrosion. Several benefits have been counted [15], including:

- Operating cost savings of 38%–42%
- Dry equipment weight reduction of 32%–37%
- Operating equipment weight reduction of 34%–40%
- Total dry weight reduction of 44%–47%
- Total operating weight reduction of 44%–50%
- Footprint requirement reduced by 40%–50%
- Capital cost reduction of 35%–40%

The advantages of membrane contactors mainly point toward high selectivity, membrane modularity, and compactness [10]. Their main disadvantage is enhanced mass transfer resistance, mostly when membranes are wetted. However, reduced mass transfer coefficients can be compensated by the numerous advantages (e.g., significantly increased interfacial area), which make membrane contactors intensely modest with conventional systems in CO_2 capture.

The performance of different amine solutions for CO_2 and H_2S removal has been investigated in a microporous symmetric polypropylene hollow fiber module. The simultaneous absorption of two gases (20% H_2S and 17% CO_2) from air was studied. The H_2S/CO_2 selectivity was found to be over 30 for tryethanolamine (TEA), 11 for 2-amino,2-methyl-1-propanol (AMP), and 5 for 2-(ethylamino)-ethnol (EAE) [7].

The separation of CO_2 from CH_4 by means of a system where a carrier solution (liquid membrane) is supplied to the feed side (at high pressure) and permeated through a UF membrane (polyethersulfone)—the surface of which is covered with a thin layer of the membrane liquid at the low-pressure side—was studied. Helium was used as a sweep gas and an aqueous solution of DEA was chosen as carrier for the experiments. The CO_2 permeance increased by increasing the carrier solution circulation rate due to the convective flow and the CO_2/CH_4 selectivity was found to be 1,970 [17].

9.2.3 SO$_2$ and Mercury Removal

Acid rain is now one of the major air pollution problems. Usually, air pollution control devices have been fixed and have worked to prevent the emission of pollutants. However, due to high cost, alternative equipment to replace the expensive devices is anticipated. SO_2 is one of the major sources that cause acid rain and support the reactions that create ozone depletion in the stratosphere [18].

As a result, many countries have implemented strict regulations regarding SO_2 emissions from coal- and oil-fired boilers in power plants, which are one of the main sources of these emissions. The SO_2 content in the flue gas is usually below about 0.1%–0.4% by volume [19]. But, the volume of the gas formed worldwide is so large that a significant amount of SO_2 seems to have been introduced into the atmosphere [20].

Gas absorption using a membrane gas–liquid contactor is a developing technology for selective separation of gaseous components, especially hydrophobic microporous membranes, which have attracted growing attention recently as contactors in chemical absorption to remove CO_2, H_2S, and SO_2 from flue and natural gases [21,22], as well as in the membrane distillation process for desalination [23,24] (see Figure 9.3).

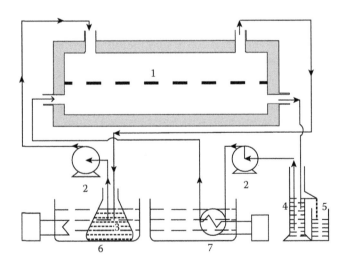

FIGURE 9.3

A schematic of a direct contact membrane distillation setup. (1) Hydrophobic porous membrane, (2) electromagnetic pump, (3) feed tank, (4) permeate tank, (5) acceptor, (6) homoeothermic system, and (7) refrigeration equipment.

Generally, removal of sulfur dioxide from gas mixtures by selective absorption into a liquid is a common method to separate and concentrate sulfur dioxide for industrial uses [25]. Regenerative processes have been industrialized in order to increase the raw materials' efficiency [26–30]. However, the use of scrubbers (technology based on dispersive absorption) leads to solvent losses due to drop dragging, depending on the contact and solvent. Recently, the performance of supported ionic liquid membranes [31–35] and polymerizable room temperature ionic liquids as gas separation membranes has been reported [36–38]. The use of suitable membrane equipment—such as a hollow fiber module where gas and liquid flow on opposite sides of the membrane and a fluid–fluid interface forms inside each membrane pore with a very small pressure drop across the membrane—has also been introduced. Mass transfer takes place by diffusion across the interface and the driving force for the separation is a concentration gradient based on solubility. Therefore, it has been claimed that the membrane-based gas–liquid procedure is a combined process merging the traditional technique of gas absorption into solvents and a membrane contactor as a mass transfer device [39,40].

The advantage of adopting this technique is its controlled interfacial area and independent control of gas and liquid flow rates, avoiding drop dragging.

A ceramic hollow fiber contactor has been used to investigate SO_2 removal from a gas stream with a typical composition of roasting processes including CO_2 (10 vol%), air, and SO_2 in the range of 0.15–4.8 vol%. N,N-dimethylaniline has been chosen as solvent in order to study the absorption process because it is frequently used in the roasting industry for SO_2 recovery. An ionic liquid, 1-ethyl-3-methylimidazolium ethyl sulfate, has been used as absorption solvent in order to avoid solvent losses due to solvent volatilization. The overall mass transfer coefficients have been found, presenting a higher resistance to mass transfer when the ionic liquid is used. Gas and liquid mass transfer coefficients have been estimated, and their contribution to the overall resistance indicated that the membrane resistance is the main contribution to be taken into consideration. The description was based on the wetting fraction, which means the ratio of the pore length wetted by liquid to the total length, taking a value of around 74% and 4% when N,N,dimethylaniline and 1-ethyl-3-methylimidazolium ethyl sulfate are used, respectively [41].

Mercury is a hazardous chemical that exists in a variety of gas streams. Emissions of mercury to the atmosphere occur during several industrial activities, such as thermal cleaning of mercury-contaminated soil, incineration of industrial and domestic waste, processing of gas-discharge lamps, natural gas production from mercury-containing wells, etc.

An oxidative gas absorption technique has been used to remove free metallic mercury vapor. H_2O_2 and $K_2S_2O_8$ have been chosen as oxidizing agents for carrying out the tests. PTFE membranes have been selected for the entire test on the basis of stability. The study investigated whether the liquid flow rate affected the mass transfer coefficient only when working at low oxidation potential. The regeneration of the absorbent liquid was obtained by the precipitation of mercury sulfide [42].

9.3 Liquid Stream Treatment

9.3.1 Wastewater Treatment

Porous ceramic membranes can offer a healthier and enduring alternative to polymeric membranes in aqueous microfiltration (MF) and ultrafiltration (UF) processes. Due to their extraordinary chemical, thermal, and mechanical properties, they not only can be operated under harsh conditions, but also can be cleaned by a backwashing method or by using strong cleaning agents; they can be sterilized at high temperatures, providing dependable and consistent performance over long periods of time. These traditional ceramic membranes are fabricated by multiple layer deposition steps on top of a membrane matrix followed by sintering (heat treatment via several steps) to achieve the desired final selectivity for MF and UF membranes.

Possible applications of ceramic membranes for water treatment include treatment of municipal and industrial wastewater, the production of drinking water, treatment of produced water, and use in the food and beverage industries. Effective execution of ceramic membranes in these industries has been achieved with stable and long-term operation, but the high investment cost of ceramic membranes remains the main restriction to large-scale water treatment [43].

In recent research, a ceramic MF membrane was successfully fabricated by a uniaxial pressing method and sintered at 950°C using natural bentonite (SiO_2 and Al_2O_3) for economic treatment of industrial wastewater. The porosity of the membrane was found to be 32.12% and the average pore size of the membrane was 1.70 μm. The prepared membrane offered good mechanical strength (22 MPa flexural strength) and good chemical stability in harsh media (both in acidic and alkaline media). After 2 h of filtration, the removal percentage of suspended matter for the two effluents A and B was between 94% and 99%, leading to complete discoloration of textile wastewater [44].

An application of a cross-flow MF process as a pretreatment step, followed by cross-flow UF and nanofiltration (NF) systems as final-treatment techniques for efficient treatment of oilfield-produced water and different model solutions using ceramic membranes was reported in the recent past (Figure 9.4). The fabricated ceramic membrane systems under micro- (0.1 μm), ultra- (0.05 μm), and nanofiltration (1000 Da) conditions proved to be economically attractive for the removal of oil from used model solutions and oilfield-produced water. Data showed that the MF membrane individually was able to remove 45% oil from wastewater—better than that of the UF (~28%). But,

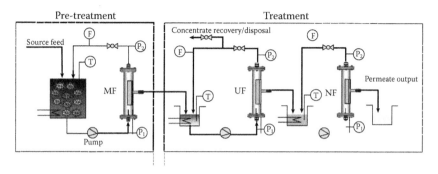

FIGURE 9.4
Schematic diagram of the laboratory scale cross-flow filtration system (with pretreatment and/ or final treatment) and liquid circulation.

in combination of MF with UF, almost 73% oil can be removed, whereas only the NF membrane alone can remove 58% of oil in the produced water [45].

In another approach, a new porous tubular membrane was fabricated based on mineral coal fly-ash for the treatment of dying effluents generated by the washing baths in the textile industry. The membrane was fabricated in two different steps: Finely ground mineral coal powder was calcined at 700°C for 3 h. The mesoporous layer was then cast by the slip-casting method using a suspension made of the mixture of fly-ash powder, water, and polyvinyl alcohol (PVA). The optimized membrane (30% PVA, 66% water, and 4% fly-ash powder) was defect free with 51% porosity, 20 μm thickness, and 0.25 μm mean pore. The performance of the fly-ash microfiltration membrane was excellent and the permeate flux was obtained about 100 L h^{-1}m^{-2} [46].

In the past, a new method of combining cross-flow microfiltration, flocculation, and ceramic membranes with a dead-end filtration system for the treatment of dyeing effluent containing DSD (4,4′-diamino stilbene-2,2′ disulfonic acid) acidic wastewater, sulfur black wastewater, and 2,3-acidic waste water [47] was successfully attempted. The membrane was mounted over a rotary disk to minimize the concentration polarization and fouling. After the experiment, the conclusions were that

- The higher the disk rotary rate is, the higher the filtration rate of sulfur black wastewater will be. When disk rotary rate is at 800 rpm, the filtration rate of DSD acidic wastewater is reduced.
- At operation pressure at 0.1 MPa, the filtration rate of wastewater is the highest.
- Dynamic filtration rate is higher than dead-end filtration rate.
- Compared with organic membrane, ceramic membrane has exceptional performance of cleaning and regeneration. The results show that the dynamic filter equipped with a ceramic plate membrane as the filtration membrane has great advantages.

The treatment of wastewater not only is restricted to conventional ceramic membranes but also reveals the possibilities of development and use of advanced ceramic techniques such as membrane distillation (MD).

Direct contact membrane distillation (DCMD) and solvent extraction (SX) have been applied in series to recover water and acid from acidic mining waste solutions. SX is a well-established treatment method to purify and recover metals from waste solutions. Experiments showed that the concentration of H_2SO_4 increased from 0.85 M in the feed solution to 4.44 M in the concentrate in the DCMD step with the synthetic acidic waste solution. Sulfate and metal separation efficiency was found to be >99.99% and the overall water recovery exceeded 80%. After recovery of water with DCMD, the concentrated solution was then subjected to recovery of sulfuric acid using SX with an organic system containing 50% TEHA and 10% ShellSol A150 (mixture of C 9-11 hydrocarbons with >99% aromatic content) in octanol. It was observed that over 80% H_2SO_4 was extracted in a single contact from the waste solution containing 245 g/L H_2SO_4 and metals with various concentrations. After three stages of successive extraction, nearly 99% of acid had been extracted [48].

DCMD has also been applied along with vacuum membrane distillation (VMD) configurations to obtain a purified stream to reuse from polyphenol-enriched olive mill wastewater (OMWW). VMD is an attractive and a cost-competitive membrane separation technology. In VMD, the vacuum pressure is applied to the permeate side of the membrane and it is maintained at just less than the saturation pressure of the volatile solvent to be separated from the hot feed solution. Experiments have been carried out on capillary membranes using a membrane module equipped with three commercial polypropylene membranes (area is around 30 cm²) with pore size of 0.2 μm, thickness of 0.4 mm, and inner diameter of 1.8 mm. DCMD tests have confirmed rejections of reducing species (polyphenols) of around 99.9% or the three temperatures investigated (30°C, 40°C, 50°C), and the flux at 50°C was obtained as 6.5 kg m⁻² h⁻¹. During VMD tests, a rejection of 99.6% polyphenols was obtained, while the permeate flux was achieved around 19 kg m⁻² h⁻¹ at 50°C [49].

9.3.2 Fruit Juice Clarification

Clarification is a vital step in the processing of fruit juice, primarily in order to remove pectin and other carbohydrates present in the juice. Usually, clarifying procedures can be achieved by various traditional techniques, such as centrifugation, enzymatic treatment, or applying clarifying agents such as gelatin bentonite, silica sol, and polyvinyl pyrrolidone. But, these processes are labor intensive, time consuming, and not continuously operated.

Membrane processes are systems used in numerous production sectors today, since this separation process involves no phase change or chemical agents. The introduction of these technologies in the manufacture of fruit juices provides an additive-free juice with high quality and a natural, fresh taste. Juice clarification, stabilization, depectinization, and concentration are typical steps in which membrane processes such as MF, UF, NF, and reverse osmosis have been successfully applied [50–53]. In the same way, clarifications based on membrane processes, mainly UF and MF, have replaced conventional fining, resulting in the elimination of the use of fining agents and a simplified process for continuous production. But the problem in using membrane for clarifying juices is the decay of permeate flux due to fouling.

A low-cost ceramic membrane for juice clarification (fresh mosambi juice) using a uniaxial dry compaction method was fabricated. The performance of the optimized membrane for mosambi juice was highly satisfactory. The optimized membrane provided a membrane flux of 90 to 44×10^{-6} $m^3/m^2 \cdot s$ at 206.7 kPa, with a permeate juice quality of negligible pectin content (expressed in terms of alcohol insoluble solids [AISs]) content for the enzyme-treated centrifuged juice (ETCJ) [54].

A low-cost ceramic MF membrane has also been applied successfully for clarification of mosambi juice. The proposed low-cost ceramic membranes yielded to provide higher membrane fluxes, adequate product quality, and lower membrane fouling. Typical permeate fluxes were observed to vary from 5.78×10^{-6} to 13.45×10^{-6} $m^3/m^2 \cdot s$ for centrifuged mosambi juice (CJ) and 14.07×10^{-6} to 60.64×10^{-6} $m^3/m^2 \cdot s$ for ETCJ at 82.7 kPa (differential pressure) for different membranes [55].

In a recent study, orange juice foulant particles were recognized and the hydraulic reversibility of the membrane fouling was investigated using a suitable lab-scale dead-end microfiltration unit with cross-flow mode. The influence of membrane average pore size, juice particle size distribution, and shear forces on filtration tests in dead-end filtration cells was discovered through a D-optimal experimental design. The results of this design showed that an anticipation of orange juice fouling tendencies, at pilot-scale microfiltration, is possible when a stirred dead-end filtration test is carried out by the ceramic membrane with an average pore size of 0.2 µm, juice fractionation prior to filtration (500 g/1 min), and rotational speed of the blade of 1000 rpm [56].

Low-cost fly-ash-based inorganic membranes have been used for the microfiltration of kiwifruit juice. These membranes have high porosities and small membrane intrinsic resistances and, as a result, provide high membrane fluxes. The optimized microfiltration membrane performed excellently in removing bacteria from the starting juice, resulting in a clean product [57].

Clarification of red raspberry (*Rubus idaeus*) juice has been investigated for cross-flow membrane filtration (MF or UF). A loss of anthocyanins in fresh and reconstituted clear juice is much smaller when the juice has been clarified using MF (pore size of 0.2 µm) than by the conventional flocculation with gelatin and bentonite. After MF through multichannel ceramic membrane, the residual pectin is completely removed [58].

These studies further support and confirm the greater applicability of low-cost ceramic membranes for juice- and beverage-processing applications in contrast to the widely applicable polymeric membranes.

9.3.3 Heavy Metal Separation

The recovery of heavy metal contaminants in liquid streams is an important target to keep our environment clean, and much research has been devoted to the development of selective extraction systems in order to meet the strict governmental limits.

Mercury (Hg) is one of the heavy metal contaminants of most concern, and coal combustion is an important emission source for atmospheric mercury. A porous ceramic membrane doped with transition metals such as Mn, Mo, and Ru (as MnO_2, MoO_3, and RuO_2) is fabricated and worked as a reactor in order to understand the performance of membrane delivery and catalytic oxidation (MDCO) at low temperature, conversion efficiency of Hg^0, efficiency of oxidants, influence of catalyst composition, and different components of flue gas. The results show that the conversion efficiencies of Hg^0 with a modified Mo–Ru–Mn catalyst are good with modified catalysts in the presence of 8 ppmv HCl [59].

Lead ion (Pb (II)) is one of the highly toxic heavy metal ions in the environment that has adverse effects on both human health and atmosphere. In a very recent study, amino-functionalized metal–organic frameworks (MOFs) were coupled with ceramic membrane ultrafiltration (CUF) for adsorptive removal of Pb (II) from wastewater. The results showed that the MOFs could be completely retained (100%) by the membrane with an average pore size of 50 nm. The maximum removal (61.4%) of Pb (II), lowest flux decline, and lowest degree of membrane fouling were observed at TMP = 0.15 MPa, CFV = 4.0 ms^{-1}, and T = 35°C, respectively. In addition, the adsorption capacity at equilibrium was found to be remarkably high (1795.3 mg.g^{-1}). The adsorbed Pb (II) could be successfully desorbed by controlling pH at 4.5 during six cycles. After each experiment, the membrane was cleaned using 0.5% (w/w) ammonium citrate followed by 0.5% (w/w) HNO_3 solution; each cleaning lasted for an hour at 18°C–50°C. Therefore, MOFs combined with ceramic membrane ultrafiltration have great potential in heavy metal wastewater treatment [60].

Arsenic is a semimetallic toxic and carcinogenic element. It can be found in groundwater under natural conditions as a result of volcanic emissions, weathering reactions, or biological activity. As(III) is much more harmful than As(V) and also much more difficult to remove from water. Thus, the removal of arsenic from drinking water streams using membrane-based technology has attracted great research attention over the past decades.

In one approach, a hydrophobic NF membrane has been synthesized, characterized, and employed as distributor for the efficient ozonation of water and hence the oxidation of As(III)–As(V). Fe_3O_4 nanoparticles have been employed as adsorbers for the removal of As(V) from drinking water streams.

The Fe_3O_4 nanoparticles are synthesized in situ in the pores of ultrafiltration membranes. It is reported that the proposed hybrid membrane process is able to remove As(III) and As(V) completely from drinking water at a stable permeate flux of ~75 L $min^{-1}m^{-2}$ under a pressure difference of $3 \times 10^5\,Nm^{-2}$. The membrane has been regenerated at 110°C and no effluent streams with detectable arsenic levels have been found as arsenic accumulates inside the membrane pores [61].

9.3.4 Aroma Compound Recovery

Aroma compound recovery is an application in the food industry for effluent treatment and for controlling food flavors.

The extraction of aroma compounds through a Liqui-Cel hollow fiber contactor from aqueous feeds to sunflower oil has been analyzed for different aroma compounds, such as 2-butanone, 2-hexanone, 2-heptanone, and 2-nonanone. The entire process is done using two different configurations: feed at shell side and oil at the lumen of fibers and feed inside the fibers and oil at the shell side. It has been observed that higher mass transfer efficiency has been achieved when water is fed at the shell side [62].

In another approach, nondispersive solvent extraction of sulfur compounds was carried out with a Liqui-Cel hollow fiber contactor. The investigated aroma compounds were dimethyldisulfide, dimethyltrysulfide, and S-methyl thiobutanoate. The extractant yield for all aroma compounds was 90%–99% and the obtained fluxes were high [63].

9.4 Fuel Cell Applications

Ceramic materials have played a vital role in shaping the ethos of prehistoric civilizations. In history, ceramic products have been developed using the same basic methods (i.e., the excavation of earthy clay, mixing with water, creating into shapes, and drying under the sunlight before finally roasting in fire).

In the past, ceramic membranes have been progressively used in a wide range of industrial applications that include the biotechnological, pharmaceutical, dairy, food, and beverage industries, as well as the chemical and petrochemical industries, microelectronics, metal finishing, and power generation. Ceramics are still used for the electrochemical treatment of wastewater, mainly by transforming pollutants into nontoxic materials. In addition, ceramics have also been employed in high-temperature fuel cells for many years.

In 1839, the principles of fuel cell operation were first described by Sir William Grove, who used hydrogen and oxygen as the reactants [64]. The use of ceramics in fuel cells was first reported in 1937, when a ceramic solid-oxide fuel cell (SOFC) was operated [65]. SOFCs consist of three ceramic layers: two electrodes with an electrolyte in the middle. Ceramic is the material of choice for SOFCs because it offers high thermal stability and provides a useful electrode material because both its porosity and permeability can be modified.

In order to fight the challenge of climate change, renewable energy technologies need to be improved and optimized. Biotransformation schemes can help in handling these challenges, and microbial fuel cells (MFCs) are one such technology that can be specifically advantageous for low-grade waste (Figure 9.5). MFCs utilize electroactive bacteria, which generate electricity by consuming organic pollutants.

The use of ceramic materials as part of the MFC was first reported by a group of researchers in 2003 [66]. This study demonstrated a proton permeable porcelain separator placed between graphite electrodes. Protons are able to pass through the microporous system utilizing separate metal catalysts at each electrode (Mn^{2+} at the anode and Fe at the cathode). In another study, a similar setup was described using a porcelain membrane combined with a solar cell, which operated stably for a longer period [67].

Later, another work from the same group reported that ceramic earthen pots at different pH can be utilized while producing power from real (rice mill) wastewater [68]. This work showed a higher power using nonsustainable chemicals. Another investigation claimed that power could be produced from locally sourced materials including ceramics (terracotta flower pot), salt, and hay as the feedstock [69]. Similarly, clayware pots with ceramics, including the deposition of salts on the cathode, were also used [70].

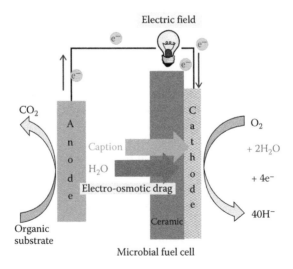

FIGURE 9.5
Schematic highlighting the mechanism of electro-osmotic drag in ceramic microbial fuel cells.

Research has shown that this novel approach is considerably less expensive. It can provide a natural, stable environment for the bacteria, while also empowering a more efficient system for energy harvesting. It is envisioned that, in the near future, ceramic MFC systems may be entirely fabricated through low-cost modified manufacturing for a wide range of target applications.

9.5 Other Applications

Process intensification consisting of the development of novel devices and techniques such as membrane contactors is expected to bring dramatic improvements in manufacturing and processing compared to those commonly used these days. Membrane contactors offer cheaper and sustainable technologies due to substantial decrease in equipment size/production capacity ratio, energy consumption, or waste production.

An alumina-based tubular ceramic membrane contactor (MWCO; 1000 kDa) has been used to recover iodine from brine water via oxidation of iodide by ozone [71]. The membrane contactor contains 19 channels inside its configuration. The experiments were carried out with variation in pH, gas–liquid flow rate ratio, and membrane length. After the experiment, the conclusions were that

- The best arrangement for the iodide oxidation process is the acidic iodide solution passing through the shell side and ozone (gas) passing through the tube side.
- The application of a membrane contactor is confirmed to improve the iodide oxidation process.
- The highest mass transfer coefficient for this system is 1.88×10^{-6} m/s at pH of 1, gas–liquid flow rate ratio of 35, and membrane length of 0.5 m.

An alternative to reducing environmental effects and costs associated to solvent losses that has been considered is a ceramic hollow fiber membrane contactor as the membrane device to recover sulfur dioxide from gas streams. From experimental results, Luis et al. have concluded that the economic feasibility of the investigated membrane device depends on the environmental limitations because the higher the inlet SO_2 concentrations are, the higher is the process efficacy that has to be attained to reach the specified outlet concentration. Also, if the required process efficiency is higher than 80%, the investment cost of the investigated membrane device is too high and not competitive on an industrial scale [72].

In a recent study, a superhydrophobic ceramic membrane contactor was used to capture CO_2 from coal-fired power plants. The membrane contactor has been fabricated from an alumina tube with a ZrO_2 layer by means of grafting with fluoroalkylsilane (FAS) in a triethoxy-1H,1H,2H,2H-tridecafluoro-*n*-octylsilane solution. A high CO_2 removal efficiency (>90%) was achieved [73].

A ceramic membrane contactor with a zirconia separation layer (pore size: 3 nm) has been modified by immobilizing an ionic liquid (IL) on the zirconia porous surface. The membrane became catalytically active for oxidation of CO by controlled growth of Au nanoparticles, which served as the active phase supported on the IL-incorporating porous membrane walls. The CO oxidation reactions in the membrane contactor were performed in the flow-through configuration. Efficient catalytic oxidation of CO has been attained at temperatures above 100°C [74].

9.6 Commercial Applications

The principal use today of membrane at the commercial level is related to water purification systems. Production of drinking water using ceramic membrane was introduced commercially by Metawater Co., Ltd, RWB, in 2006 in Almelo, the Netherlands (Figure 9.6) [75]. The company's product has excellent features, such as

- No membrane breakage/during 100% integrity
- Long service life/10 years—100% guarantee
- Robustness against raw water fluctuations
- High recovery rate and easy handling of backwash waste
- Less malfunction and breakdown
- Low power consumption and low operations and maintenance cost
- Excellent chemical resistance
- High reduction of bacteria and viruses

Pall's Membralox® ceramic membrane elements are ideal for applications that include extreme processes, such as high solids bulk processes and the use of high temperatures or pressures or aggressive solvents, where significant long-term durability is required [76]. Membralox membrane elements are available in three different channel diameters to allow optimization for loading solids. Some configurations are available with exclusive longitudinal

Andijk 3, pilot phase 1 (2006)

Andijk 3, pilot phase 2 (2009) C-19

Andijk 3, pilot phase 3 (2012) C-200

Andijk 3, 2014 (The Netherlands)

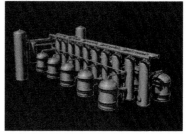

FIGURE 9.6
Drinking water applications by Metawater Co., Ltd, RWB, Almelo, The Netherlands.

permeability gradients; this aids in the control of permeate rate along the length of the module. Membralox ceramic membranes contain a highly controlled surface membrane layer that is formed on the inner (feed-side) surface of a more open support layer. Different modes of materials are available: ultrapure, alumina (MF), zirconia, and titania (UF). They offer excellent features of ceramic membranes (Figure 9.7):

- Reliability
- Ease of use
- High flux
- Proven long operational life
- Wide chemical stability (pH: 14)
- Excellent thermal stability
- Can be sterilized
- Ability to withstand high-frequency back-pulsing cycles

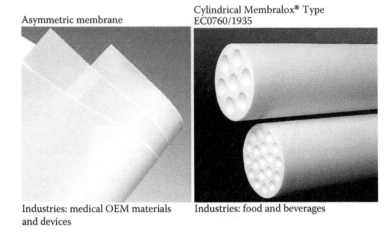

Asymmetric membrane

Cylindrical Membralox® Type
EC0760/1935

Industries: medical OEM materials
and devices

Industries: food and beverages

FIGURE 9.7
Membranes and materials by PALL Corporation, NY.

9.7 Summary

In this chapter, we have described various applications of ceramic membranes and membrane contactors. The examples of separation of different pollutants are summarized in the following:

- Gaseous stream treatment
  - VOC removal—novel techniques applied
  - Acid gas removal—use of membrane contactors is suitable and advantageous
  - SO_2 and mercury removal—hollow fiber contactors, ionic liquid membranes
- Liquid stream treatment
  - Wastewater treatment—use of membrane contactors is appreciable.
  - Fruit juice clarification—low-cost ceramic membranes are good.
  - Heavy metal separation—conventional ceramic membranes can be used.
  - Aroma compounds recovery—hollow fiber contactors are suitable.
- Fuel cell technology—challenging and under investigation
- Other applications—membrane contactors offering less expensive, sustainable technologies
- Commercial applications—well appreciable but not commercialized broadly; improvement required

References

1. J.J. Cai, K. Hawboldt, M.A. Abdi, Improving gas absorption efficiency using a novel dual membrane contactor. *Journal of Membrane Science*, 510 (2016) 249–258.
2. S.K. Padhi, S. Gokhale, Biological oxidation of gaseous VOCs—Rotating biological contactor a promising and eco-friendly technique. *Journal of Environmental Chemical Engineering*, 2 (4) (2014) 2085–2102.
3. T.K. Poddar, S. Majumder, K.K. Sirkar, Removal of VOCs from waste gas streams by permeation in a hollow fiber permeator. *Journal of Membrane Science*, 120 (1996) 221–237.
4. T.K. Poddar, K.K. Sirkar, A hybrid of vapor permeation and membrane-based absorption-stripping for VOC removal and recovery from gaseous emissions. *Journal of Membrane Science*, 132 (1997) 229–233.
5. J. Kujawa, S. Cerneaux, W. Kujawski, Highly hydrophobic ceramic membranes applied to the removal of volatile organic compounds in pervaporation. *Chemical Engineering Journal*, 260 (2015) 43–54.
6. Z. Qi, E.L. Cussler, Microporous hollow fibers for gas absorption: I. Mass transfer in the liquid. *Journal of Membrane Science*, 23 (1985) 321–332.
7. Z. Qi, E.L. Cussler, Microporous hollow fibers for gas absorption: II. Mass transfer across the membrane. *Journal of Membrane Science*, 23 (1985) 333–345.
8. H.H. Park, B.R. Deshwal, I.W. Kim, H.K. Lee, Absorption of SO_2 from flue gas using PVDF hollow fiber membranes in a gas–liquid contactor. *Journal of Membrane Science*, 319 (2008) 29–37.
9. K. Li, D. Wang, C.C. Koe, W.K. Teo, Use of asymmetric hollow fibre modules for elimination of H_2S from gas streams via a membrane absorption method. *Chemical Engineering Science*, 53 (1998) 1111–1119.
10. A. Gabelman, S.-T. Hwang, Hollow fiber membrane contactors. *Journal of Membrane Science*, 159 (1999) 61–106.
11. J.-L. Li, B.-H. Chen, Review of CO_2 absorption using chemical solvents in hollow fiber membrane contactors. *Separation & Purification Technology*, 41 (2005) 109–122.
12. A. Mansourizadeh, A.F. Ismail, Hollow fiber gas–liquid membrane contactors for acid gas capture: A review. *Journal of Hazardous Materials*, 171 (2009) 38–53.
13. Z. Cui, D. deMontigny, Part 7: A review of CO_2 capture using hollow fiber membrane contactors. *Carbon Management*, 4 (2013) 69–89.
14. E. Drioli, A.I. Stankiewicz, F. Macedonio, Membrane engineering in process intensification—An overview. *Journal of Membrane Science*, 380 (2011) 1–8.
15. O. Falk-Pedersen, M.S. Grønvold, P. Nøkleby, F. Bjerve, H.F. Svendsen, CO_2 capture with membrane contactors. *International Journal of Green Energy*, 2 (2005) 157–165.
16. D. deMontigny, P. Tontiwachwuthikul, A. Chakma, Comparing the absorption performance of packed columns and membrane contactors. *Industrial and Engineering Chemistry Research*, 44 (2005) 5726–5732.
17. M. Teramoto, N. Takeuchi, T. Maki, H. Matsuyama, Facilitated transport of CO_2 through liquid membrane accompanied by permeation of carrier solution. *Separation & Purification Technology*, 27 (2002) 25–31.
18. S. Ebrahimi, C. Picioreanu, R. Kleerebezem, J.J. Heijnen, M.C.M. van Loosdecht, Rate-based modeling of SO_2 absorption into aqueous $NaHCO_3/Na_2CO_3$ solutions accompanied by the desorption of CO_2. *Chemical Engineering Science*, 58 (2003) 3589–3600.

19. S.B. Han, A study on the absorption rate of SO_2 into aqueous solution by single rising bubble through a quiescent column, PhD dissertation, Department of Chemical Engineering, Pusan University, Korea, 1986.

20. C.S. Chang, G.T. Rochell, SO_2 absorption into aqueous solutions. *AIChE Journal,* 27 (1981) 292–297.

21. A. Gabelman, S.T. Hwang, Hollow fiber membrane contactors. *Journal of Membrane Science,* 159 (1999) 61–106.

22. J.L. Li, B.H. Chen, Review of CO_2 absorption using chemical solvents in hollow fiber membrane contactors. *Separation and Purification Technology,* 41 (2005) 109–122.

23. C. Feng, R. Wang, B. Shi, G. Li, Y. Wu, Factors affecting pore structure and performance of poly(vinylidene fluoride-co-hexafluoro propylene) asymmetric porous membrane. *Journal of Membrane Science,* 277 (2006) 55–64.

24. S. Karoor, K.K. Sirkar, Gas absorption studies in microporous hollow fiber membrane modules. *Industrial & Engineering Chemistry Research,* 32 (1993) 674–684.

25. H. Müller, Sulfur dioxide, in *Ullmann's encyclopaedia of industrial chemistry,* Wiley-VCH Verlag GmbH & Co., Weinheim, Germany, 2005.

26. E. Bekassy-Molnar, E. Marki, J.G. Majeed, Sulphur dioxide absorption in air-lift tube absorbers by sodium citrate buffer solution. *Chemical Engineering and Processing: Process Intensification,* 44 (9) (2005) 1039–1046.

27. C.M. Yon, G.R. Atwood, Jr., C.D. Swaim, Use UCAP for sulphur recovery. *Hydrocarbon Processing,* 58 (1979) 197–200.

28. D.W. Neumann, S. Lynn, Kinetics of the reaction of hydrogen sulfide and sulfur dioxide in organic solvents. *Industrial & Engineering Chemistry Process Design and Development,* 25 (1986) 248–251.

29. M.H.H. van Dam, A.S. Lamine, D. Roizard, P. Lochon, C. Roizard, Selective sulfur dioxide removal using organic solvents. *Industrial & Engineering Chemistry Research,* 36 (1997) 4628–4837.

30. H. Li, D. Liu, F. Wang, Solubility of dilute SO_2 in dimethyl sulfoxide. *Journal of Chemical & Engineering Data,* 47 (2002) 772–775.

31. P. Scovazzo, J. Kieft, D.A. Finan, C. Koval, D. DuBois, R. Noble, Gas separations using non-hexafluorophosphate [PF6]-anion supported ionic liquid membranes. *Journal of Membrane Science,* 238 (2004) 57–63.

32. J. Ilconich, C. Myers, H. Pennline, D. Luebke, Experimental investigation of the permeability and selectivity of supported ionic liquid membranes for CO_2/He separation at temperatures up to 125°C. *Journal of Membrane Science,* 298 (2007) 41–47.

33. Y. Jiang, Z. Zhou, Z. Jiao, L. Li, Y.Wu, Z. Zhang, SO_2 gas separation using supported ionic liquid membranes. *Journal of Physical Chemistry B,* 111 (2007) 5058–5061.

34. S. Hanioka, T. Maruyama, T. Sotani, M. Teramoto, H. Matsuyama, K. Nakashima, M. Hanaki, F. Kubota, M. Goto, CO_2 separation facilitated by task-specific ionic liquids using a supported liquid membrane. *Journal of Membrane Science,* 314 (2008) 1–4.

35. P. Luis, L.A. Neves, C.A.M. Afonso, I.M. Coelhoso, J.G. Crespo, A. Garea, A. Irabien, Facilitated transport of CO_2 and SO_2 through supported ionic liquid membranes (SILMs). *Desalination,* 245 (1) (2009) 485–493.

36. J.E. Bara, S. Lessmann, C.J. Gabriel, E.S. Hatakeyama, R.D. Noble, D.L. Gin, Synthesis and performance of polymerizable room-temperature ionic liquids as gas separation membranes. *Journal of Membrane Science*, 46 (2007) 5397–5404.

37. J.E. Bara, C.J. Gabriel, E.S. Hatakeyama, T.K. Carlisle, S. Lessmann, R.D. Noble, D.L. Gin, Improving CO_2 selectivity in polymerized room-temperature ionic liquid gas separation membranes through incorporation of polar substituents. *Journal of Membrane Science*, 321 (2008) 3–7.

38. J.E. Bara, E.S. Hatakeyama, C.J. Gabriel, X. Zeng, S. Lessmann, D.L. Gin, R.D. Noble, Synthesis and light gas separations in cross-linked gemini room temperature ionic liquid polymer membranes. *Journal of Membrane Science*, 316 (2008) 186–191.

39. A.E. Jansen, R. Klaassen, P.H.M. Feron, J.H. Hanemaaijer, B. Meulen, Membrane gas absorption processes in environmental applications, in *Membrane processes in separation and purification*, Kluwer Academic Publishers, Amsterdam, the Netherlands, 1994, pp. 343–356.

40. S.B. Iversen, V.K. Bhatia, K. Dam-Johansen, G. Jonsson, Characterization of microporous membranes for use in membrane contactors. *Journal of Membrane Science*, 130 (1997) 205–217.

41. P. Luis, A. Garea, A. Irabien, Zero solvent emission process for sulfur dioxide recovery using a membrane contactor and ionic liquids. *Journal of Membrane Science*, 330 (2009) 80–89.

42. R. van der Vaart, J. Akkerhuis, P. Feron, B. Jansen, Removal of mercury from gas streams by oxidative membrane gas absorption. *Journal of Membrane Science*, 187 (2001) 151–157.

43. M. Lee, Z. Wu, K. Li, Advances in ceramic membranes for water treatment, in *Advances in membrane technologies for water treatment*, Woodhead Publishing, Sawston, Cambridge, England (2015), pp. 43–82.

44. A. Bouazizi, S. Saja, B. Achiou, M. Ouammou, J.I. Calvo, A. Aaddane, S.A. Younssi, Elaboration and characterization of a new flat ceramic MF membrane made from natural Moroccan bentonite. Application to treatment of industrial wastewater. *Applied Clay Science*, 132–133 (2016) 33–40.

45. M. Ebrahimi, K. Shams Ashaghi, L. Engel, D. Willershausen, P. Mund, P. Bolduan, P. Czermak, Characterization and application of different ceramic membranes for the oil-field produced water treatment. *Desalination*, 245 (2009) 533–540.

46. I. Jedidi, S. Saïdi, S. Khemakhem, A. Larbot, N.E. Ammar, A. Fourati, A. Charfi, A.B. Salah, R.B. Amar, Elaboration of new ceramic microfiltration membranes from mineral coal fly ash applied to waste water treatment. *Journal of Hazardous Materials*, 172 (2009) 152–158.

47. L. Xu, W. Li, S. Lu, Z. Wang, Q. Zhu, Y. Ling, Treating dyeing waste water by ceramic membrane in cross-flow microfiltration. *Desalination*, 149 (2002) 199–203.

48. U.K. Kesieme, H. Aral, Application of membrane distillation and solvent extraction for water and acid recovery from acidic mining waste and process solutions. *Journal of Environmental Chemical Engineering*, 3 (2015) 2050–2056.

49. M.C. Carnevale, E. Gnisci, J. Hilal, A. Criscuoli, Direct contact and vacuum membrane distillation application for the olive mill wastewater treatment. *Separation and Purification Technology*, 169 (2016) 121–127.

50. L.R. Fukomoto, P. Delaquis, B. Girard, Microfiltration and ultrafiltration ceramic membranes for apple juice clarification. *Journal of Food Engineering,* 63 (1998) 845–850.

51. V. Gokmen, Z. Borneman, H.H. Nijhuis, Improved ultrafiltration for color reduction and stabilization of apple juice. *Journal of Food Engineering,* 63 (1998) 504–507.

52. F. Vaillant, A. Millan, M. Dornier, M. Decloux, M. Reynes, Strategy for economical optimization of the clarification of pulpy fruit juices using cross-flow microfiltration. *Journal of Food Engineering,* 48 (2001) 83–90.

53. A. Cassano, L. Donato, E. Drioli, Ultrafiltration of kiwifruit juice: Operating parameters, juice quality and membrane fouling. *Journal of Food Engineering,* 79 (2007) 612–621.

54. S. Emani, R. Uppaluri, M.K. Purkait, Preparation and characterization of low cost ceramic membranes for mosambi juice clarification. *Desalination,* 317 (2013) 32–40.

55. B.K. Nandi, R. Uppaluri, M.K. Purkait, Identification of optimal membrane morphological parameters during microfiltration of mosambi juice using low cost ceramic membranes. *LWT—Food Science and Technology,* 44 (2011) 214–223.

56. D. Layal, W. Christelle, R. Julien, K.-G. André, D. Manuel, D. Michèle, Development of an original lab-scale filtration strategy for the prediction of microfiltration performance: Application to orange juice clarification. *Separation and Purification Technology,* 156 (2015) 42–50.

57. G. Qin, X. Lu, W. Wei, J. Li, R. Cui, S. Hu, Microfiltration of kiwifruit juice and fouling mechanism using fly-ash-based ceramic membranes. *Food and Bioproducts Processing,* 96 (2015) 278–284.

58. G.T. Vladisavljevic, P. Vukosavljevic, M.S. Veljovic, Clarification of red raspberry juice using microfiltration with gas backwashing: A viable strategy to maximize permeate flux and minimize a loss of anthocyanins. *Food and Bioproducts Processing,* 91 (2013) 473–480.

59. Y. Guo, N. Yan, S. Yang, P. Liu, J. Wang, J. Qu, J. Jia, Conversion of elemental mercury with a novel membrane catalytic system at low temperature. *Journal of Hazardous Materials,* 213–214 (2012) 62–70.

60. N. Yin, K. Wang, L. Wang, Z. Li, Amino-functionalized MOFs combining ceramic membrane ultrafiltration for Pb (II) removal. *Chemical Engineering Journal,* 306 (2016) 619–628.

61. S. Sklari, A. Pagana, L. Nalbandian, V. Zaspalis, Ceramic membrane materials and process for the removal of As(III)/As(V) ions from water. *Journal of Water Process Engineering,* 5 (2015) 42–47.

62. A. Baudot, J. Floury, H.E. Smorenburg, Liquid–liquid extraction of aroma compounds with hollow fiber contactor. *AIChE Journal,* 47 (2001) 1780–1793.

63. F.X. Pierre, I. Souchon, M. Marin, Recovery of sulfur aroma compounds using membrane based solvent extraction. *Journal of Membrane Science,* 187 (2001) 239–253.

64. N.Q. Minh, T. Takahashi, *Science and technology of ceramic fuel cells,* Elsevier, Amsterdam, the Netherlands (1995).

65. E. Baur, H. Preis, Uber Brennstoff-Ketten mit Festleitern. *Zeitschrift für Elektrochemie,* 43 (1937) 727–732.

66. D.H. Park, J.G. Zeikus, Improved fuel cell and electrode designs for producing electricity from microbial degradation. *Biotechnology and Bioengineering,* 81 (3) (2003) 348–355.

67. H.N. Seo, W.J. Lee, T.S. Hwang, D.H. Park, Electricity generation coupled with wastewater treatment using a microbial fuel cell composed of a modified cathode with a ceramic membrane and cellulose acetate film. *Journal of Microbiology and Biotechnology,* 19 (9) (2009) 1019–1027.

68. M. Behera, P.S. Jana, T.T. More, M.M. Ghangrekar, Rice mill wastewater treatment in microbial fuel cells fabricated using proton exchange membrane and earthen pot at different pH. *Bioelectrochemistry,* 79 (2) (2010) 228–233.

69. F.F. Ajayi, P.R. Weigele, A terracotta bio-battery. *Bioresource Technology,* 116 (2012) 86–91.

70. P. Chatterjee, M.M. Ghangrekar, Design of clayware separator-electrode assembly for treatment of wastewater in microbial fuel cells. *Applied Biochemistry and Biotechnology,* 173 (2014) 378–390.

71. I.G. Wenten, H. Julian, N.T. Panjaitan, Ozonation through ceramic membrane contactor for iodide oxidation during iodine recovery from brine water. *Desalination,* 306 (2012) 29–34.

72. P. Luis, A. Garea, A. Irabien, Environmental and economic evaluation of SO_2 recovery in a ceramic hollow fibre membrane contactor. *Chemical Engineering and Processing,* 52 (2012) 151–154.

73. X. Au, L. An, J. Yang, S.-T. Tu, J. Yan, CO_2 capture using a super hydrophobic ceramic membrane contactor. *Journal of Membrane Science,* 496 (2015) 1–12.

74. A.V. Perdikaki, A.I. Labropoulos, E. Siranidi, I. Karatasios, N. Kanellopoulos, N. Boukos, P. Falaras, G.N. Karanikolos, G.E. Romanos, Efficient CO oxidation in an ionic liquid-modified, Au nanoparticle loaded membrane contactor. *Chemical Engineering Journal,* 305 (2016) 79–91.

75. Metawater Co. Ltd., Almelo, the Netherlands, http://www.rwbalmelo.nl /(accessed November 11, 2016).

76. Pall Corporation, New York, Membralox® Ceramic Membrane Products, http://www.pall.co.in (accessed November 11, 2016).

Nomenclature

AACVD	aerosol-assisted chemical vapor deposition
AFM	atomic force microscopy
AGMD	air gap membrane distillation
AMP	2-amino, 2-methyl-1-propanol
ANOVA	analysis of variance
APCVD	atmospheric pressure chemical vapor deposition
ASTM	American Society for Testing and Materials
BA	boric acid
CCD	central composite design
CMC	ceramic matrix composite
CMR	catalytic membrane reactor
CNMR	catalytic nonpermselective membrane reactor
CSTR	continuous stirred tank reactor
CVD	chemical vapor deposition
CVI	chemical vapor infiltration
CTAB	cetyl trimethyl-ammonium bromide
DCMD	direct contact membrane distillation
DGM	dusty gas model
DLICVD	direct liquid injection chemical vapor deposition
DLS	dynamic light scattering
DOE	design of experiments
DSC	differential scanning calorimetry
EAE	2-(ethylamino)-ethnol
EDF	equilibrium deposition filtration
EDTA	ethylene-diamine-tetra-acetic acid
EDX	energy dispersive x-ray
ELS	electrophoretic light scattering
EME	electromembrane extraction
ESF	elelctrosinter forging
ESR	electron spin resonance
EtAc	ethyl acetate
ETCJ	enzyme-treated centrifugal juice
FAST DHT	fast direct hot pressing
FBCMR	fluidized bed catalytic membrane reactor
FBMR	fluidized bed membrane reactor
FEG-SEM	field emission gun scanning electron microscopy
FESEM	field emission scanning electron microscopy
FTIR	fourier transform infrared spectroscopy
GAM	gas absorption membrane
GISAX	grazing incidence small-angle x-ray scattering

HIP	hot isostatic pressing
ICDD	International Center for Diffraction Data
IEP	isoelectric point
IMR	inert membrane reactor
LPCVD	low-pressure chemical vapor deposition
LPSA	laser particle size analyzer
LRS	Laser Raman spectroscopy
MARS	membrane aromatic recovery system
MBR	membrane bioreactor
MCr	membrane crystallization
MD	metal dispersion (%)
MDCO	membrane delivery and catalytic oxidation
MEA	membrane electrode assembly
MF	microfiltration
MFC	microbial fuel cell
MOF	metal–organic framework
MPCVD	microwave plasma chemical vapor deposition
MR	membrane reactor
MSA	metallic surface area ($m^2.g^{-1}$)
MSR	methanol steam re-forming
MTBE	methyl-*tert*-butyl-ether
MWCO	molecular weight cutoff
NF	nanofiltration
ODH	oxydehydrogenation
OMW	olive mill wastewater
PBCMR	packed-bed catalytic membrane reactor
PBMR	packed-bed membrane reactor
PE	polyethylene
PECVD	plasma-enhanced chemical vapor deposition
PEG	polyethylene glycol
PEO	polyethylene oxide
PFR	plug flow reactor
PMMA	polymethyl methacrylate
PP	polypropylene
PS	polystyrene
PSPD	position-sensitive photo diode
PTFE	polytetrafluoroethylene
PV	pervaporation
PVA	polyvinyl alcohol
PVD	physical vapor deposition
QD	quantum dot
RBAO	reaction bonding of alumina
RBC	rotating biological contactor
RBSN	reaction bonded silicon nitride
RHP	rapid hot pressing

RO	reverse osmosis
RPCVD	remote plasma chemical vapor deposition
RSM	response surface methodology
SAED	selected-area electron diffraction
SANS	small-angle neutron scattering
SAXS	small-angle x-ray scattering
SCR	selective catalytic reduction
SDS	sodium dodecyl sulfate
SGMD	sweep gas membrane distillation
SLM	supported liquid membrane
SM	sodium metasilicate
SOFC	solid oxide fuel cell
SPS	spark plasma sintering
SWAX	small- and wide-angle x-ray scattering
SX	solvent extraction
TAA	tetra alkyl ammonium
TEA	triethanolamine
TEHA	tris-2-ethylhexylamine
TF	turnover frequency (cm^{-1})
TGA	thermogravimetric analysis
TOC	total organic carbon
TPR	temperature programmed reduction
UF	ultrafiltration
UHVCVD	ultrahigh vacuum chemical vapor deposition
VMD	vacuum membrane distillation
VOC	volatile organic compound
XPS	x-ray photoelectron spectroscopy
XRD	x-ray diffractometer

List of Symbols

A	Model term for preparation pressure
A_1	Slope
A_{eff}	Effective membrane area (m²)
A_f	Arrhenius constant for feed side (mol cm⁻³ s⁻¹ atm⁻¹·⁵)
A_i	Area of peak I (m²)
A_r	Arrhenius constant for product side (mol cm⁻³ s⁻¹ atm⁻¹·⁷⁵)
B	Model term for sodium metasilicate (SM) content
B_0	A specific membrane parameter (m²)
B_1	Intercept
C	Model term for boric acid (BA) content
C_A	Concentration of reactant gas after time t (mol.m⁻³)
C_{A_0}	Initial concentration of reactant gas (mol.m⁻³)
C_f	Solute concentration in feed (mol.m⁻³)
$C_{overall_0}$	Overall initial concentration of both the reactants (mol.m⁻³)
C_p	Solute concentration in permeate (mol.m⁻³)
$C_{R_{inlet}}$	Initial concentration of reactant gases in the inlet of the reactor (mol.m⁻³)
$C_{R_{outlet}}$	Concentration of feed gases present in the permeate (mol.m⁻³)
C_s	Reactant concentration at the external surface of catalyst (mol.m⁻³)
c_t	Total concentration of the reactant gases (mol.m⁻³)
d	Collision diameter of the gas molecules (m)
d_i	Partial desirability function for specific responses
d_i	Pore diameter of the ith pore (μm)
D_{ij}	Effective diffusivity of the reactant gases (m².min⁻¹)
$D_{inert,k}/D_K$	Knudsen diffusion coefficient (m².min⁻¹)
d_p	Pore diameter (m)
d_s	Equivalent particle diameters (m)
d_{sl}	Density of solid–lubricant solution (g.m⁻³)
$dX_{overall}$	Change in rate of disappearance of the reactant gases with time (mol.m⁻³.min⁻¹)
E_{af}	Activation energy in feed side (kcal mol⁻¹)
E_{ar}	Activation energy in product side (kcal mol⁻¹)
h	Height of the membrane (m)
J_i	Molar flux of ith component gases (mol.m⁻².min⁻¹)
J_j	Molar flux of jth component gases (mol.m⁻².min⁻¹)
J_K	Molar flux (mol.m⁻².min⁻¹)
$J_{membrane,i}$	Molar flux of the reactant gases through the membrane wall (mol.m⁻².min⁻¹)

J_{visc}	Molar flux considering transport mechanism, viscous flow (mol.m^{-2}.min^{-1})
J_w	Volumetric flux of the solvent (m^3.m^{-2}.s^{-1})
k	Effective permeability factor (m.min^{-1})
k_b	Backward rate constant (mol. L^{-1}.min^{-1})
K_{eq}	Equilibrium constant
k_f	Forward reaction rate constant (mol. L^{-1}.min^{-1})
k_{ij}	Mass transfer coefficient for components i and j (m.min^{-1})
l	Pore length (m)
L or l_m	Thickness of the membrane (m)
l_d	Distance between the reactor wall and the membrane surface where the diffusion occurs (m)
L_p	Liquid permeability (mol.m^{-2}.min^{-1}.MPa)
m	Mass of the coating over the exterior surface of the membrane (g)
$m.w$	Molecular weight of the metal (g.mol^{-1})
M_A, M_B	Molar mass of the reactant gases (g.mol^{-1})
m_s	Mass of the sample (g)
M_{N_2}	Molecular weight of the inert gas (g.mol^{-1})
$\%M$	Percent metal
n	Number of pores (%)
N_A/N	Avogadro's number
p	Pressure at which reactant gases are entered into the reactor (MPa)
P	Pressure applied to force mercury into a circular cross-section capillary of pore diameter
p_1	Downstream pressure (Pa)
P_2	Membrane pressure at permeate side (MPa)
p/p_o	Relative pressure (Pa)
P_{amb}	Ambient pressure (MPa)
p_h	Upstream pressure (Pa)
p_i	Partial pressure of the species i (MPa)
P_i	Permeance (mol.m^{-2}.min^{-1}.MPa)
P_{std}	Standard pressure (MPa)
P_{tot}	Total pressure (MPa)
$\bar{p}$	Average pressure across the membrane (Pa)
ΔP	Transmembrane pressure drop (MPa)
ΔP_{start}	Breakthrough pressure (Pa)
Q	Volumetric flow rate (m^3.min^{-1})
r	Diameter of the particle (m)
R	Universal gas constant (m^3.MPa.K^{-1}.mol^{-1})
R_1	Rejection (%)
$R1$	Model response for membrane flexural strength
R^2	Coefficient of multiple determination

$R2$	Model response for membrane porosity
R^2_{Adj}	Adjusted statistic coefficient
r_a	Reaction rate per volume of the catalyst (mol.m^{-3}.min^{-1})
R_b	Rate of the reaction in the bulk (mol.m^{-3}.min^{-1})
r_g	Average pore radius (m)
R_{int}	Rate of the reaction at the interface (mol.m^{-3}.min^{-1})
r_o	Outer radius of the membrane coated with catalyst (m)
$r_{overall}$	Rate of disappearance of the reactant gases (mol.m^{-3}.min^{-1})
r_p	Pore radius (m)
R_p	Catalyst particle radius (m)
$r_{p.max}$	Largest pore radius of the membrane (m)
S	Permeable area of the membrane (m^2)
S_f	Stoichiometric factor
T	Temperature (K,°C)
t	Time required to convert reactant gases into products (min)
T_0	Phase transition temperature of the liquid (K)
T_{amb}	Ambient temperature (K,°C)
T_c	Thickness of the coated layer (m)
T_{std}	Standard temperature (K,°C)
ΔT	Difference in temperature (K)
V	Reactor volume (m^3)
V or u	Molecular mean velocity of the operating gas (m.min^{-1})
v_A, v_B	Diffusion volumes of the reactant gases (m^3)
V_{ads}	Amount of gas adsorbed (mL)
V_c	Volume of the coated surface (m^3)
V_g	Molar volume of the gas at standard temperature and pressure (mL)
v_i	Stoichiometric coefficient of component i
V_{inj}	Volume of active gas injected (mL)
v_j	Stoichiometric coefficient of component j
V_m	Volume of the fabricated membrane (m^3)
V_{mol}	Molar volume of the condensable vapor (m^3.mol^{-1})
V_{syr}	Syringe volume injected (mL)
v_ϕ	Volumetric flow rate (m^3.min^{-1})
W_1	Weight of the dry membrane (m)
W_2	Weight of the wet membrane (m)
x_i	Molar fraction of the species i
x_j, x_j^{int}	Molar fractions at the bulk and the interface
$x_{S_8}, x_{H_2O}, x_{H_2S}$ and x_{SO_2}	Mole fractions of products and reactants
X	Conversion of the reactants into product (%)
$\bar{x}$	Average mole fraction between the bulk and the catalyst–membrane surface interface
Δx	Difference in mole fraction between the bulk and the catalyst–membrane surface interface

Greek Letters

η	Viscosity of the gas (MPa.min)
α	Axial points in CCD
ε	Porosity
τ	Tortuosity factor
ε/τ^2	Effective porosity
ρ_{H_2O}	Density of water (g.m^{-3})
φ_{W-P}	Weisz–Prater criterion
δ_m	Cross-sectional area of active metal atom (m^2)
λ	Mean free path
λ_1	Wavelength of x-rays
Θ	Geometric factor associated with tortuosity factor
μ	Viscosity (Pa.S)
γ	Surface tension of mercury (N.m^{-1})
γ_s	Liquid–solid interfacial tension (J.m^{-2})
γ_{SV}	Solid–vapor interfacial energy
γ_{SL}	Solid–liquid interfacial energy
γ_{LV}	Liquid–vapor interfacial energy
θ	Angle of contact of mercury on the material being imposed
θ_C	Equilibrium contact angle
ζ	Process parameter

Index

Printed and bound by CPI Group (UK) Ltd, Croydon, CR0 4YY

24/10/2024

01778301-0007